ROCKETS AND MISSILES

ROCKETS AND MISSILES

THE LIFE STORY OF A TECHNOLOGY

A. Bowdoin Van Riper

GREENWOOD TECHNOGRAPHIES

GREENWOOD PRESS
Westport, Connecticut • London

Library of Congress Cataloging-in-Publication Data

Van Riper, A. Bowdoin.
    Rockets and missiles : the life story of a technology / A. Bowdoin Van Riper.
       p.   cm.—(Greenwood technographies, ISSN 1549–7321)
    Includes bibliographical references and index.
    ISBN 0–313–32795–5 (alk. paper)
    1. Rocketry (Aeronautics)—History.   2. Ballistic missiles—History.   I. Title.
  II. Series.
  TL781.V36   2004
  621.43'56—dc22   2004053045

British Library Cataloguing in Publication Data is available.

Library of Congress Catalog Card Number: 2004053045
ISBN: 0–313–32795–5
ISSN: 1549–7321

First published in 2004

Greenwood Press, 88 Post Road West, Westport, CT 06881
An imprint of Greenwood Publishing Group, Inc.
www.greenwood.com

Printed in the United States of America

The paper used in this book complies with the
Permanent Paper Standard issued by the National
Information Standards Organization (Z39.48–1984).

10 9 8 7 6 5 4 3 2 1

For Janice P. Van Riper

who let a starstruck kid stay up long past his bedtime
to watch Neil Armstrong take "one small step"

Contents

viii   Contents

Series Foreword

In today's world, technology plays an integral role in the daily life of people of all ages. It affects where we live, how we work, how we interact with each other, and what we aspire to accomplish. To help students and the general public better understand how technology and society interact, Greenwood has developed *Greenwood Technographies*, a series of short, accessible books that trace the histories of these technologies while documenting *how* these technologies have become so vital to our lives.

Each volume of the *Greenwood Technographies* series tells the biography or "life story" of a particularly important technology. Each life story traces the technology from its "ancestors" (or antecedent technologies), through its early years (either its invention or development) and rise to prominence, to its final decline, obsolescence, or ubiquity. Just as a good biography combines an analysis of an individual's personal life with a description of the subject's impact on the broader world, each volume in the *Greenwood Technographies* series combines a discussion of technical developments with a description of the technology's effect on the broader fabric of society and culture—and vice versa. The technologies covered in the series run the gamut from those that have been around for centuries—firearms and the printed book, for example—to recent inventions that have rapidly taken over the modern world, such as electronics and the computer.

While the emphasis is on a factual discussion of the development of the technology, these books are also fun to read. The history of technology is full of fascinating tales that both entertain and illuminate. The authors—all experts in their fields—make the life story of technology come alive, while also providing readers with a profound understanding of the relationship of science, technology, and society.

Acknowledgments

The greatest debt I have incurred in writing this book is, unfortunately, one that I cannot fully acknowledge. *Rockets and Missiles* is a brief introduction to a sprawling topic. It is, therefore, a synthesis of knowledge gathered and conclusions drawn by others. It is also part of a series that, by design, uses formal references only to acknowledge direct quotations. I have turned, in the course of researching and writing *Rockets and Missiles,* to the published work of countless other scholars. The books, articles, and web sites on which I relied most heavily are noted—along with selected others—in the Further Reading section at the end of this book. A heartfelt "thank you" is all I can offer to the authors of the rest.

Other debts are, happily, more easily acknowledged.

The staff of the Lawrence V. Johnson Library of Southern Polytechnic State University was cheerfully unflappable in the face of my frequent requests for obscure materials. Leigh Hall of the Interlibrary Loan Department deserves special thanks. The staffs of the Cobb County Public Library System and the libraries of Emory University and the Georgia Institute of Technology also aided my search for "just the right material" on countless rocket-related topics.

Many times in the course of this project I benefited from informal input from fellow scholars. I am especially grateful—for references suggested, opinions critiqued, experience shared, and misunderstandings clarified—to

Amy Foster, Slava Guerevich, John Krige, Roger Launius, Frederick Ordway, Asif Siddiqi, and Kristen Starr.

Kevin Downing, who conceived the *Greenwood Technographies* series and invited me to be part of it, played a critical role in the initial shaping of this book. The scope, organization, and approach of *Rockets and Missiles* all reflect his valuable input. He is, to paraphrase Gilbert and Sullivan, "the very model of a modern scholarly editor," and it has been a pleasure to work with him. The editorial and production staffs at Greenwood Press have been both friendly and efficient—a combination of which every writer dreams.

Finally, as always, I owe a considerable debt to Julie, Joe, and Katie. Julie cheerfully accepted the domestic chaos that comes with writing a book and, once again, gave me reason to reflect on the joys of being married to a fellow historian of science and technology. Joe and his friends, though they may not have known it, became the faces on the "target audience" I kept in mind as I wrote. Katie buoyed me with her enthusiasm and, with a single sentence—"You're writing a book about *rockets*?!"—made it all seem worthwhile.

Timeline

Before 1100 CE	Gunpowder invented in China.
Mid-1100s	Probable first use of military rockets by the Chinese.
Mid-1200s	Probable first use of military rockets by Arabs and Europeans.
Around 1300	First detailed, written descriptions of rocket technology appear in European and Arabic manuscripts.
Around 1400	Rocket technology known and used in all major Eurasian civilizations.
Mid-1600s	Rockets appear in military manuals such as *Artis Magnae Artilliae* (Europe) and *Wu Pei Chih* (China).
1780–1799	Indian troops use rockets against the British in the Mysore Wars.
1805	William Congreve adapts Indian rocket technology for European use.
1805–1815	British troops use Congreve rockets against Danes, French, and Americans.
1820s	Congreve-type rockets adapted for use in lifesaving and whaling.
1840	William Hale invents the first spin-stabilized (stickless) rocket.

1870s	Black powder rockets begin to decline as "tube" artillery improves.
1903	Konstantin Tsiolkovsky describes multistage, liquid-propellant rockets in his paper "Exploring Space with Reactive Devices."
1919	Robert Goddard, unaware of Tsiolkovsky's work, describes liquid-propellant rockets in "On a Method of Reaching Extreme Altitudes."
1923	Hermann Oberth, unaware of Tsiolkovsky or Goddard, describes liquid-propellant rockets in *Die Rakete zu den Planetenraum* (The Rocket into Planetary Space).
1926	Goddard builds and flies the first liquid-propellant rocket.
1925–1935	Amateur rocket societies founded in the United States, USSR, and Germany.
Late 1930s	Civilian engineers backed by government funding invent storable liquid propellants, improved solid propellants, and other key technologies.
1939–1945	World War II: revival of rocket artillery, first use of guided missiles.
Late 1945	Soviet and American rocket engineers, assisted by German émigrés, begin building and testing high-altitude research rockets.
1947	Rocket-powered Bell X-1 exceeds the speed of sound in level flight.
1956	First warplane armed only with missiles becomes operational.
1957	Launch of *Sputnik I*, the first artificial satellite to orbit the Earth.
1957–1958	First IRBMs become operational.
1958	First successful use of air-to-air missiles in combat (Taiwan versus China).
1958–1960	First ICBMs become operational.
	China acquires ballistic missile technology from the USSR.
1960	First missile-carrying submarines become operational.
1961	Launch of *Vostok I*, the first manned spacecraft to orbit the Earth.
	First successful launch of a silo-based missile.
1962	Cuban Missile Crisis brings United States and USSR close to nuclear war.

1964	First successful test of an antiballistic missile (by the United States).
1964–1973	Vietnam War: first sustained use of surface-to-air, air-to-air, and air-to-surface guided missiles in combat.
1967 (Apr)	First manned test flight of the *Soyuz* spacecraft.
1967 (Nov)	First test flight of the *Saturn V*, the largest space launch vehicle ever built.
1969 (Jan)	First manned test flight of the complete *Apollo* spacecraft.
1969 (Jul)	*Eagle*, an Apollo LM, makes the first manned landing on the moon.
1970	United States develops its first MIRVs.
1970s	USSR exports FROG and Scud missiles throughout the Middle East.
1971	First test flight of the *DF-5*, China's first ICBM.
1972 (Jan)	President Richard Nixon announces the space shuttle program.
1972 (May)	Soviet and American leaders sign the SALT and ABM treaties.
1972 (Dec)	Final lunar landing mission of Project Apollo.
1973 (Jan)	USSR abandons its manned lunar landing program.
1973 (May)	First flight of the *Soyuz-U* launch vehicle.
1975	USSR develops its first MIRVs.
1979	First flight of the European Space Agency's *Ariane* launch vehicle.
1980–1988	Iran-Iraq War: first use of missiles against cities since World War II.
1981	First orbital test flight of the space shuttle.
1981–1983	Small-scale conflicts (United States–Libya, Britain-Argentina, Israel-Syria) reveal the now-dominant role of tactical missiles in modern warfare.
1983	President Ronald Reagan announces the Strategic Defense Initiative.
1985	Reykjavik Summit: first serious proposals for missile-*reduction* treaties between the United States and USSR.
1986	Loss of the space shuttle *Challenger*.
1987	Soviet and American leaders sign the INF treaty.
	Missile Technology Control Regime established.

1988	Only test flight of the Energia heavy booster and Buran space-plane.
1990s	China, India, Pakistan, Iran, and North Korea expand missile programs.
1991	Persian Gulf War: first use of modern cruise missiles (Tomahawk) and antiballistic missiles (Patriot) in combat.
1998	First operational ion thruster tested aboard *Deep Space 1* robot spacecraft.
2001–2002	President George W. Bush announces U.S. withdrawal from the ABM Treaty and plans to deploy a national missile defense system.
2003 (Feb)	Loss of the space shuttle *Columbia*.
2003 (Oct)	China launches its first manned spaceflight.
2004	President George W. Bush announces plans to establish a permanent U.S. lunar base and send U.S. astronauts to Mars.

1

Introduction

◆

"Rocket science" has become verbal shorthand for complexity. Saying that something is "not rocket science" suggests that it is simple and easily grasped. The expression, in turn, says something about the way we think of rockets: fantastically complex, unimaginably powerful, the highest of high technology.

Electricity, airplanes, computers, and nuclear weapons have all been thought of that way at some point in the twentieth century, but none of them is today. Electricity and airplanes have grown familiar. We take the electrical appliances in our home or office for granted, and think of long-distance trips as exotic only when we take a vehicle other than an airplane. Computers have grown domesticated. Tall, boxy metal cabinets have given way to sculptured plastic cases, and modern operating systems allow users to manipulate brightly colored pictures rather than typing an exotic human-machine language in glowing green characters across a forbiddingly empty black screen. Nuclear weapons have grown distant. The last above-ground nuclear explosions in the United States (visible to eyewitnesses and cameras) took place more than forty years ago, and the terrorist's low-tech "dirty bomb" or vial of anthrax now seems a greater threat. Only rockets remain awe-inspiringly powerful, stunningly complex, *and* regularly visible in Hollywood fiction and CNN broadcasts from Cape Canaveral.

Rocket technology is nearly 1,000 years old, having been invented (almost certainly in China) sometime between 1000 and 1150. It was, within a century or two of its invention, known to all the major urban civilizations of Eurasia: Chinese, Indian, Muslim, and European. Though used in war to bombard enemies and in peace to send messages, carry ropes, and hunt whales, rockets rarely changed the course of events and never altered the fabric of everyday life. Midway through the twentieth century, however, they began to do both. The change was a result of two parallel revolutions: a technological one that transformed the way rockets worked, and a conceptual one that transformed the way first engineers and later the general public thought about them. Those dual revolutions set the stage for three new applications of rocket technology: long-range "strategic" missiles, short-range "tactical" missiles, and "launch vehicles" to carry payloads into space. Rockets to fulfill each of those three roles had been built and flown by the late 1940s. Over the next six decades, refined versions of those rockets have reshaped our world, transforming science, politics, economic, and, above all, warfare.

This book is a history of rocket technology from its first days to our present day: a history of how and why rockets changed over time, and of what those changes meant.

ROCKET SCIENCE 101

Everyday usage aside, the basic principles of rocketry are not "rocket science." A rocket is simply a machine that exploits Newton's third law of motion. It propels itself forward by "throwing" a steady stream of matter out behind it.

Applying a force to an object causes it to accelerate: that is, to change its speed, the direction of its motion, or both. Newton's third law states that forces come in matched pairs. "To every reaction force," in other words, "there is an equal [in magnitude] and opposite [in direction] reaction force." Say, for example, that it takes 20 pounds of force to close the driver's door of a car. When someone pushes it just hard enough to close it, they are applying 20 pounds of force to the door and the door is (simultaneously) applying 20 pounds of force to them. The third law is deeply counterintuitive because, under normal circumstances, people are conscious of the forces they apply to objects they move, but not of the forces that the objects apply to them. Trying to close a car door while standing on icy pavement in smooth-soled shoes makes the paired forces more apparent. The force applied to the door accelerates it enough to make it swing closed. Simultaneously, however, the force that the door applies to the person closing it accelerates them enough to make their feet skid across the ice.

The same principle of paired forces is at work when someone inflates a balloon and then releases it without tying the neck shut. The tightly stretched skin of the balloon immediately begins to contract back to its natural, much smaller shape. The contracting balloon exerts a force on the air molecules inside, accelerating them out through the open neck in a steady stream. At the same time, the air molecules exert a force on the balloon that accelerates *it* across the room in the opposite direction. The balloon's flight is "powered" by the acceleration given to the air molecules by the contraction of the skin. Standard rubber balloons can, therefore, fly only for the second or two it takes their skins to return to its normal shape. Balloons made of silvery plastic like Mylar, whose walls are nonelastic, cannot fly in this way at all.

A rocket exploits Newton's third law in the same way that a deflating balloon does. A force pushes a steady stream of gas out behind the rocket, and a force of equal magnitude pushes the rocket itself in the opposite direction: forward. The critical difference between a rocket and a deflating balloon—the difference between a child's toy and a world-changing tool—is a matter more of engineering than of science. The balloon is "fueled" with air that is blown into it, held momentarily by clamping the neck shut, and then released when the neck is opened. The rocket is fueled with combustible chemicals that, when burned inside the rocket, yield a cloud of hot gas. All gas (whether exhaust gas in a rocket or air in a balloon) expands to fill its container. Hotter gasses expand more rapidly than colder ones, since their molecules are moving faster. The burning of a rocket's propellant steadily adds more and more hot gas to the confined space inside the rocket, raising the pressure that the gas exerts. The pressure forces a steady stream of gas out through the open vent (or vents) at the rear of the rocket: the exhaust plume whose acceleration in one direction causes the rocket to accelerate in the opposite direction.

The force produced by a rocket is called "thrust," and is usually measured in pounds or kilograms. The most critical measures of a rocket's performance are tied directly to the amount of thrust it produces. The "specific impulse" of a rocket is the amount of thrust produced by 1 pound of propellant in 1 second—a measure of the fuel's potency and the engine's efficiency. The "thrust-to-weight ratio" is exactly what its name suggests: a comparison between the thrust that a rocket produces and its weight. The higher the thrust-to-weight ratio, the greater the rocket's ability to carry a payload: weight above and beyond the systems and propellant required to make the rocket function. Payload capacity, range, and altitude—measurements of a rocket's ability to do useful work in the real world—are all intimately linked to the thrust it produces.

Producing thrust has, from the beginning of rocketry until today, meant burning fuel. Burning requires oxygen, and rockets carry both a supply of fuel and a supply of oxygen with them. What fuel and oxidizer to use is the single most important choice a rocket designer must make. Virtually everything else about a rocket's design depends, to some degree, on that choice. The first phase of this choice is whether to use liquid or solid propellants. The fuel and oxidizer in liquid-propellant rockets are carried in separate tanks, then injected into a combustion chamber by pumps or by compressed gas before being ignited. The fuel and oxidizer in solid-propellant rockets are premixed into a uniform compound that is then packed or poured (it may initially be semiliquid) into the body of the rocket. Each type of propellant has advantages and drawbacks. Liquid propellant rockets are more mechanically complex and usually heavier, but they can be "throttled," or even turned off and on, in flight. Solid propellant rockets are simpler to build and easier to store for long periods, but they are also less flexible and more difficult to "scale up" to very large sizes.

Like the plant and animal kingdoms in biology, the liquid- and solid-propellant "kingdoms" in rocket design are both highly diverse. Each encompasses a wide range of possible fuels, oxidizers, and design features. The twenty-first century promises, however, to bring even greater diversity—perhaps even in the form of entirely new "kingdoms" of propellants.

"Working fluid" is the engineering term for something that is accelerated out the back end of an engine in order to accelerate the engine forward: the air in a deflating balloon, for example. Virtually every rocket flown to date has used exhaust gasses as a working fluid. Nothing about Newton's third law, however, requires that it must be so. Rockets now being considered for use on future long-duration space missions may well abandon exhaust gasses as a working fluid and combustion as a means of generating and accelerating that fluid. They may, in other words, be the first operational rockets in history that do not actually *burn* anything. The final chapter of this book touches briefly on one such rocket.

TERMINOLOGY: ROCKETS, MISSILES, AND MOTORS

The word "rocket" was (intentionally) used throughout the preceding section without formally defining it. The subject matter did not require a formal definition, and providing one would have been complicated and needlessly distracting. Before starting the main narrative, however, a few definitions are in order.

A rocket is (and has been since the Middle Ages) a self-contained, self-propelled projectile that carries its own supplies of fuel and oxygen. The word applies equally to projectiles for military use (bombardment) and civilian use (signaling, lifesaving, fireworks). It has, since World War II, been applied only to self-propelled projectiles without onboard guidance systems. A rocket designed to be installed in a vehicle (car, aircraft, spacecraft) as a propulsion system is, technically, a "rocket motor" or a "rocket engine."

The word "missile" (for centuries, just a synonym for "projectile") now refers exclusively to a self-contained, self-propelled projectile with some form of guidance system. The first such guidance systems were developed in the 1930s, but the narrow sense of the word (shortened from "guided missile") did not come into wide use until after World War II. Germany's V-2s were missiles in the modern sense of the word, but when they were falling on London and Antwerp in 1944–1945, both sides generally referred to them as rockets. Missiles are usually powered by rocket motors, but they need not be. Cruise missiles, invented in the 1940s and widely used since 1980, are propelled wholly or partly by jet engines that use air as both an oxidizer and a working fluid.

A launch vehicle is a rocket-powered vehicle used to lift satellites and spacecraft into orbit around the Earth. Most of the launch vehicles used since the beginning of the Space Age in 1957 have been adapted from military missiles, but the term "launch vehicle" also applies to manned cargo-carrying spacecraft like the space shuttle. The term "booster rocket" is often used by the public and the mainstream press to describe unmanned launch vehicles. Professionals tend to avoid it, however, because "booster" also refers to a self-contained, solid–propellant rocket motor used to enhance the thrust of a launch vehicle or missile for specific missions.

A spacecraft is a vehicle (with or without a human crew) capable of operating beyond Earth's atmosphere. It may leave Earth under its own power (like the space shuttle) or be carried into space by a launch vehicle (like the Soyuz). It is different from a satellite because it can be steered by a human pilot or controllers on the ground. Spacecraft need not, in theory, use rocket propulsion. Plans exist for spacecraft propelled by ground-based lasers or "solar sails" designed to catch the streams of charged particles given off by the Sun. So far, however, every spacecraft to travel beyond Earth's atmosphere has done so under rocket power.

2

The Age of Black Powder, 1000–1900

The idea of blending natural ingredients into a substance that will burn or explode when ignited has been around since ancient times. One famous early example was "Greek fire," a mixture of petroleum, sulfur, resin, and pitch that sticks to almost anything and burns fiercely when ignited. Invented (as the name suggests) in Greece, it was used in the lands around the Mediterranean from the 670s CE on. Gunpowder is a simpler mixture, but a more difficult one to prepare successfully. It contains only three ingredients— charcoal, sulfur, and saltpeter (potassium nitrate)—but will work only if they are combined in the right proportions, ground into fine particles, and mixed thoroughly. Recipes for gunpowder were developed by trial and error and, once perfected, written down. The oldest surviving recipes appear in Chinese and European manuscripts from the mid-1200s CE and in Arabic manuscripts from the late 1200s. They were probably based on earlier recipes of which no copy survives, and the first successful experiments may have predated those earlier recipes by decades or even centuries.

Most historians believe that the knowledge of how to make gunpowder originated in China sometime between 500 and 900 CE, and spread westward during the 1200s. The Mongols, a nomadic people whose empire stretched from China to the plains of Hungary by the late 1200s, developed gunpowder technology after the Chinese used it against them.

They almost certainly carried it to the Arab world and may have brought it to Europe, although it is also possible that European scientists developed it independently.

The black powder that powered every rocket made before 1900 was close, but not identical, to the gunpowder poured down the barrels of early firearms. It was made of the same three components, but blended with less saltpeter and more charcoal to make it "slow": that is, to ensure that it would burn steadily rather than explosively. It was also, as rocket-makers grew more sophisticated, dampened before it was packed into the rocket's tubular body so that it would dry, in place, into a solid "cake" rather than remaining as loose granules. The technique, another Chinese invention, was originally applied to the third major gunpowder technology: fireworks.[1] Turning loose powder into cake improved its stability when stored and its reliability afterward, and fireworks—like rockets but unlike guns—were often stored for a time between being loaded with power and being fired.

The relationship between gunpowder's three offspring—guns, fireworks, and rockets—is complex, and differs from country to country and century to century. All three applications were explored, however, by each of the civilizations that acquired gunpowder in the Middle Ages. The step from gunpowder to rockets appears to have been universal.

THE MIDDLE AGES (CIRCA 1100 TO 1450)

Early guns and early rockets were variations on the same technological theme: packing gunpowder into a tube closed at one end, and then igniting the powder so that the hot gasses created by its combustion escaped from the open end. There were, however, critical differences between the two technologies. One difference is that the powder in a gun burns explosively, while the powder in a rocket burns steadily. Another is that, while a gun barrel has to withstand multiple firings, a rocket body has to withstand only one. Rocket bodies could, therefore, be built of lighter materials (wood, bamboo, or even paper) that were easier to obtain and easier to work with than the bronze or iron required for gun barrels. A third critical difference is that the combustion gasses in a gun push a projectile out of the tube (barrel),

---

1. "Fireworks," then as now, include a wide range of explosive devices: firecrackers that explode on the ground, fixed tubes that spew colored smoke or fire, bombs thrown into the air by short-barreled cannons, and rockets that rise into the sky under their own power. Up to 1900, fireworks rockets were essentially military rockets with decorative rather than destructive "warheads." This chapter will, therefore, treat them only in passing.

while those in a rocket push the tube (body) itself. Rocket designers did not, therefore, face a problem that bedeviled gun designers: how to make a projectile that fits into the barrel loosely enough to move freely but tightly enough to trap combustion gasses behind it. Gun making was (and still is) a skill that can be acquired only by working with someone who already understands the process. The specialized knowledge necessary to build a rocket—what size to make the tube, how to prepare the powder, and how much powder to use—could be efficiently summarized in words and diagrams. The ability to put it on paper meant that knowledge of rocket building was spreading rapidly by the end of the Middle Ages.

The earliest surviving "books" to mention rockets were produced in the twelfth century. They were not books in the modern sense, but bound manuscripts. The printing press did not yet exist, and they had to be copied by hand if they were to be duplicated, which limited how many could be produced and how widely they could be distributed. Whether they are the oldest such works, or only the oldest to survive, is an open question.

The earliest references to rockets in print tend to be vague. Chinese manuscripts from the early 1200s, for example, refer to weapons they call "firelances" and "fire arrows," and suggest that those weapons were used in battle as early as 1127. Both weapons consisted of a tube of gunpowder attached to a long, pointed shaft—a description that could refer to a rocket-propelled projectile *or* to a conventional projectile designed to explode among the enemy or set fire to its target. The earliest European references to what may be rockets are equally vague. The Poles are said to have used them against the Tatars at the Battle of Liegnitz in 1241, and the Moors against the Spaniards in 1249 and 1288. Albertus Magnus described "flying fire"—gunpowder packed into a long, thin tube—in *De Mirabilibus Mundi* in the 1270s. Whether these descriptions reflect practical knowledge of rockets or only rumors and secondhand knowledge is far from clear.

The written record becomes clearer around 1300. *Liber Ignium,* written by an author who used the name "Marcus Graecus," was a collection of recipes for Greek fire, gunpowder, and similar substances that was compiled in the late 1200s and published around 1300. It offered two detailed formulas for rocket fuel, the first of which concluded: "Then put into a reed or hollow wood and light it. It flies away suddenly to whatever place you wish and burns up everything" (Von Braun and Ordway 1976, 3). The *Treatise on Horsemanship and Stratagems of War,* written around the same time by a Syrian named Al-Hasan al-Rammah, offers its own formulas along with more detailed descriptions of rockets and rocket-powered weapons. One such weapon was a rocket-propelled "egg that moves itself and burns," which al-Rammah states was used by Arab forces against the French at the

Battle of Damiah in 1248. The claim is important because an account of the battle by a French eyewitness also mentions an Arab weapon that slid across the ground under rocket power, spewing fire and scattering the French knights. The existence of corroborating descriptions from opposite sides of the same battle suggests that the Arabs, at least, were using military rockets by the middle of the thirteenth century. They also suggest that Al-Rammah, writing fifty years later, was describing not fanciful rumors but real military technology.

Arabic manuscripts also offer some slight corroboration of the argument that Chinese firelances and fire arrows were rockets. Shams al-din Muhammad's *Collection Combining the Various Branches of the Art*, written in the early 1300s, describes an "arrow from Cathay" that is unambiguously a rocket attached to a lance. It provides critical details such as the recipe for the powder, the method of packing it, and the placement of the fuse. "Cathay" was a medieval western name for China, and other Arab sources refer to saltpeter, the critical ingredient in gunpowder, as "snow of China." Muhammad does not say (and may not have known) when the "arrow from Cathay" arrived in the Middle East; nor does he say whether it was a straightforward copy of, or an improvement on, the Chinese original.

Wherever it originated, and however it spread, rocket technology was widely known throughout Europe and Asia by the early 1400s. Handbooks of military technology, such as Konrad Kyeser von Eichstädt's *Bellifortis* and Giovanni da Fontana's *Belliscorim Instrumentarum Liber*, discussed rockets and their applications in detail. Indian soldiers under Sultan Mahmud used rockets in their 1399 defense of Delhi against the armies of Tamerlane, Chinese armies used them against the Vietnamese near modern-day Hanoi in 1426, and French troops under Joan of Arc used them against the English at the siege of Orléans in 1428. The use of rockets as fireworks, an established tradition in China, spread throughout South Asia in the early 1400s. It reached India and the islands of Indonesia along newly opened trade routes and became a standard form of entertainment at large, public celebrations. Understanding of how rockets worked also deepened in the early 1400s. Eichstädt's *Bellifortis*, for example, noted that a rocket is pushed forward by its exhaust, and that the casing must be impervious to gas in order for the rocket to work.

THE EARLY MODERN ERA (1450–1800)

The world changed profoundly in the decades around 1450. Turkish armies captured Constantinople, erasing the last traces of the old Roman Empire and redrawing the political map of Eastern Europe and Southwest Asia. The government of China stopped sponsoring the long-distance sea voyages that had (briefly) extended its influence to Arabia and East Africa. The government of Portugal began sponsoring voyages along the coast of Africa: voyages that would bring Portuguese ships to Indian ports by the end of the century. The newly unified kingdom of Spain began to purge its territory of Muslim "infidels" whose ancestors had come by force eight centuries before. Gunpowder became common enough in Europe to bring an end to the armored knight, the stone castle, and the decentralized political system they helped to sustain. A German goldsmith named Gutenberg gave Europe the printing press, the first tool for mass-producing knowledge. In northern Italy, a once-in-a-century flowering of artistic talent set the Renaissance in motion.

Rockets played essentially the same roles in the early modern era that they had played in the Middle Ages. They were used to bombard the enemy in wartime, and to stage elaborate fireworks displays in peacetime. The rockets themselves also changed relatively little, except that they became increasingly standardized in design and construction. The rise of the printing press and the spread of printed books encouraged this standardization by making the latest information about rockets widely available. Printed descriptions of rockets and instructions on how to build them were available throughout Europe in the 1500s and early 1600s. Notable works appeared not only in traditional centers of learning like Italy and Spain, but also in still-remote areas of Europe such as Romania, Poland, and northern Germany.

A few of these works described rockets with radically new features. Even those, however, focused primarily on refining rather than transforming the state of the art. *Artis Magnae Artilliae* ("The Great Art of Artillery"), written by Polish armaments expert Casimir Siemienowicz, illustrates this contrast. Published in Amsterdam in 1650 and translated into English in 1729, it described a three-stage rocket and the use of fins to provide stability. Siemienowicz was more a technician than a visionary, however, and he also provided a formula for calculating the proper length of the guide stick: $N \times (N + 1)$, where "N" is the length of the rocket in fingers. The cumulative effect of these works was to make the best rocket-design practices of the day widely available, and so to make rockets more standardized. It is reasonable, then, to consider what a typical rocket of, say, the mid-1700s looked like (see Figure 2.1).

A black powder rocket began with a tube, generally made of cast iron if the rocket was for military use and of pasteboard (layers of paper bound together with paste) if it was for civilian use. A typical rocket tube was relatively narrow: a few inches in outside diameter, and a length five times the outside diameter. The back end of the tube was fitted with a pasteboard disk that, like a large washer, had a hole in its center. The front end was left temporarily open, so that the rocket could be filled with powder. Inserting the powder was accomplished by fitting the tube (open front end up) into a vertical wooden or metal mold. A tall cone, called the "piercer," projected upward from the bottom of the mold and entered the rocket tube through the hole in its base. The rocket-maker tamped and packed powder into the tube until it was filled to the top, then sealed the nose of the tube with

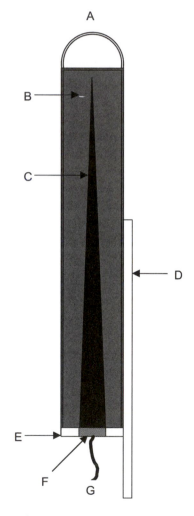

Figure 2.1: A cross-section of a generic black powder rocket from the early 1800s, showing the features common to all such rockets: warhead (A), powder (B), void space (C), guide stick (D), end cap (E), primer (F), and fuse (G). Drawn by the author.

a solid disk. The tube could then be removed from the mold, packed with powder except for a cone-shaped space where the piercer had been. French rocketeers called the space the *âme*, or "soul," and with good reason: it was essential for the rocket to operate properly. Finishing the rocket involved sealing the hole in the base with primer (finely granulated powder moistened into a paste), into which was set a fuse (a cotton string soaked in brandy or vinegar and then impregnated with more powder). A tapered wooden stick nine times or more the length of the rocket was attached to the outside of the tube at the tail end for balance.

Launching such a rocket was relatively straightforward. The rocket was laid in a V-shaped wooden trough or some other device that could be pointed and elevated, and the fuse was lit. The fire worked its way along the gunpowder-saturated string until it reached and ignited the primer. The finely ground primer, like the priming powder used in the pan of a flintlock musket, burned away in an instant, igniting the coarser powder of the main charge. The main charge in a black powder rocket burned from the inside out. The layers of powder at the edges of the empty space burned first, and the fire gradually ate its way outward toward the walls and forward toward the nose of the rocket. The hot gasses produced by the burning powder expanded in every direction, but could only escape in one: out through the hole (or "throat") left in the tail of the rocket when the primer burned away. The escaping gasses thrust the rocket forward, with the long guide stick providing balance like the tail of a kite.

Fitting black powder rockets with explosive heads—whether for war or for fireworks displays—was also relatively simple. The explosive charge, fitted with its own fuse and primer, was packed into a separate container mounted on the front of the rocket tube. The fuse passed through a small hole in the disk that sealed the front end of the rocket tube, so that the end touched the powder of the propellant charge. The powder at the front end of the tube—the last to burn, if the rocket worked properly—would therefore ignite the fuse and trigger the warhead in the last moments of the rocket's powered flight. It was a highly effective system for fireworks rockets, ensuring that detonation would occur near the peak of the rocket's trajectory. For military rockets, often fired horizontally or in shallow arcs, it was less effective. Fired at close range, they might strike their targets before the motor lit the warhead's fuse. Fired at long range, they might be destroyed by the explosion of their warhead before they reached the target. Artillerymen and ship's gunners faced similar problems when they used explosive shells in their cannons. They would not be fully solved until the invention, in the mid-nineteenth century, of chemical fuses that would detonate when they hit a solid object.

Black powder rockets were used sporadically on the battlefields of the seventeenth and eighteenth centuries. Where they *were* used, however, they tended to be used in large numbers—possibly as a way of magnifying their psychological effect and getting around their lack of accuracy. The Chinese text *Wu Pei Chih*, written in the 1620s, describes rockets with explosive warheads being fired from wooden boxes divided into cells and capable of holding 100 projectiles each. The rulers of the kingdom of Mysore, in southern India, began to equip their armies with rockets in the 1750s. Haider Ali and his son and successor, Tippoo Sahib, ultimately attached a company of rocketeers to each of their army's brigades—a total of 5,000 rocket-carrying troops by the 1790s. Their rockets, built in two standardized sizes, had tubes of cast iron rather than the then-standard bamboo or pasteboard. The use of iron added weight but also lent strength, allowing designers to make the rockets more powerful without fear that the added pressure from the expanding exhaust gasses would burst them. The extra thrust that iron tubes allowed more than compensated for the extra weight. According to Indian sources, Tippoo Sahib's rocket troops could bombard targets as much as a mile and a half away.

The military value of Indian rockets became apparent when Haider Ali and Tippoo Sahib led their troops into battle against the British army in the 1780s and 1790s. Haider Ali's victory at the Battle of Pollilur (1780), during the Second Mysore War, was due in part to rockets setting a British ammunition wagon afire. Tippoo Sahib, who ascended to the throne when his father was killed in 1782, made effective use of rockets again in his attack on the city of Travancore, which started the Third Mysore War in 1790. The final act of the Fourth Mysore War was played out in 1799 when British troops cornered Tippoo Sahib in his capital city, Seringapatam. A British force under Colonel Arthur Wellesley (later the Duke of Wellington) approached the city, but turned and fled when the Mysoreans unleashed a rocket barrage and a hail of musket fire. Ultimately, however, the British regrouped and brought their artillery to bear on the city walls. An early, lucky shot touched off a storeroom filled with rockets, and the resulting explosion opened a breach in the wall that later shots expanded. The British charged, and Tippoo Sahib died, ironically, fighting to hold a gap in his walls accidentally made by his own secret weapon.

NINETEENTH-CENTURY MILITARY ROCKETS

Tippoo Sahib's secret weapon did not remain secret for long. Word of his success with rockets reached Europe while the Mysore Wars were still going on, spurring research on military rockets in England, France, Ireland, and elsewhere. After the capture of Seringapatam and the death of Tippoo Sahib, the British shipped hundreds of rockets home to the Royal Arsenal as spoils of war. The point of the shipment was less to equip British troops with Indian rockets than to "reverse engineer" them: take them apart, study how they were made, and learn how to build rockets that were as good or better.

The comptroller of the Royal Arsenal was an old soldier named William Congreve who was also a senior officer in the Royal Artillery. His oldest son, also William, was twenty-seven when Tippoo Sahib died—a recent graduate of the University of Cambridge who practiced law, edited newspapers, and lived the high life among wealthy and titled friends in London. The younger Congreve had connections to the Royal Arsenal through his father and connections to some of the most powerful men in Britain through his friends. He also had a deep fascination with machines, and in mid-1804 he gave up both publishing and the law to pursue it. Congreve eventually received patents for things ranging from steam engines and canal locks to a new printing technique that made paper money more difficult to counterfeit. His first project, however, was to devise a weapon that could destroy the fleet of troop-carrying barges that Napoleon was assembling along the coast of France in preparation for an invasion of England. Congreve began with captured Indian war rockets and, improving on them, single-handedly brought on a revolution in rocket design.

Congreve's revolution was part of the larger Industrial Revolution that was transforming Britain in the early nineteenth century. One of the central elements of the Industrial Revolution was the standardization and mechanization of manufacturing. Products that had been made one at a time by individual workers in separate workshops were increasingly mass-produced in centralized factories. Workers who once shaped raw materials directly, using hand tools and muscle power, increasingly tended steam-powered machine tools that shaped the materials for them. Factory-made products were cheaper and more abundant than the workshop-made products that they replaced, and they were also more uniform. Even the most skilled and attentive handworker turned out products that varied slightly from one another. A well-tended machine would, in contrast, always cut a strip of fabric to the same width, plane a block of wood to the same thickness, or bore a hole to the same depth. Congreve applied this principle to rocket design. To be truly effective weapons, he concluded, rockets had to be rigidly standardized.

Congreve made three critical innovations in rocket design. The first, borrowed straight from the rocketeers of Mysore, was to use metal rather than pasteboard (or any other organic material) for the tube. The second was to use a mass-produced black powder mixed according to a standardized formula and prepared with mechanical grinding mills that produced particles of uniform size. The third was to use a device like a small pile driver—a heavy weight, lifted by ropes and pulleys and then dropped—to pack the powder into the tube. Congreve's machine-ground powder burned more smoothly than the hand-ground powders it replaced, and mechanical packing eliminated the empty or loosely packed pockets that hand packing sometimes left. His rockets developed a consistently high thrust, and their metal bodies ensured that they could withstand the increased gas pressures that produced it.

Congreve rockets thus offered not only better performance than earlier types, but more consistent performance as well. Access to the firing ranges of the Royal Arsenal allowed him to conduct extensive tests, which led to further fine tuning of both rockets and their launching apparatus. He was thus able, in 1805, to offer the Royal Army and Navy what would now be called a "weapon system": an array of rockets in various sizes, each with an appropriate launching apparatus and most with a choice of explosive or incendiary warheads.

British cannon were named, in the early nineteenth century, for the weight of the iron balls that they fired: a "9-pounder" was a relatively small gun, a "32-pounder" a relatively large one. Congreve rockets were also designated as "___-pounders," but in their case the weight was that of the largest lead ball that would fit inside the rocket tube. Those in active use ranged from 6-, 9-, 12-, and 18-pounder "light" rockets through 24- and 32-pounder "medium" rockets to 42-pounder "heavy" rockets. Tiny 3-pounders and massive 100- and 300-pounders were also developed, but the former was too small to do significant damage and the latter were too cumbersome to handle in the field.

British forces first used Congreve rockets in battle in 1805, and continued to use them throughout the wars against the French (1805–1812, 1815) and the Americans (1812–1814). A massive barrage of Congreve rockets—as many as 25,000 according to some accounts—set the city of Copenhagen, Denmark, afire in 1807, and the 150-man Royal Artillery Rocket Brigade played a critical role at the battle of Leipzig in 1813. Led by Captain Richard Bogue, it laid down a barrage that caused 2,500 French troops to break ranks and flee at a decisive moment. British rockets were also decisive at the 1814 Battle of Bladensburg in the War of 1812, which set the stage for their capture and burning of the city of Washington.

The most famous use of rockets in this war, which the British called the "Second American War," was, ironically, a failure. For nearly twenty-four hours on September 12–13, 1814, British ships anchored off Baltimore bombarded Fort McHenry with cannon and 32-pounder Congreves in an effort to force its surrender. The fort survived, but Francis Scott Key— an American envoy being held temporarily on one of the British ships— immortalized "the rockets' red glare" in his poem "The Star-Spangled Banner."

The use of Congreve rockets eventually spread well beyond Britain. They were, by the middle of the nineteenth century, in the arsenals of every major European power as well as the arsenals of the United States and a number of Middle Eastern and Latin American nations. The reasons for this wide popularity are easy to understand. Congreve rockets were a new kind of artillery that were, in many ways, superior to cannon.

Even a "light" 12-pounder Congreve had a range of a 1.25 miles— double that of contemporary light artillery. A 32-pounder could, at a range of nearly 2 miles, punch through the walls of buildings or penetrate 9 feet of earth. Rockets generated no recoil (the force that slams a cannon back when it is fired), and so could be launched from lightweight wooden frames. The frames for light rockets could be carried by individual soldiers or mounted in small oared boats; those for heavy rockets could be mounted on horse-drawn wagons and the decks of modest-sized ships. Reloading the muzzle-loading cannons used in the early nineteenth century was a complex, multistep process. Reloading a rocket frame involved little more than lifting a new rocket into position. Trained rocketeers could, as a result, fire four rounds in a minute—a pace that even the best gun crews could not match. Freed of the need to move a heavy bronze or iron cannon and its carriage, rocketeers were also more mobile than traditional artillery units. A hundred men on foot could hand-carry 10 frames and 300 light rockets to the front lines and discharge all 300 rockets in less than 10 minutes. Four horses—barely enough to pull a medium-sized cannon—could carry 4 frames and 72 rounds on their backs. Rocket troops could move fast and hit hard, a combination that endeared them to forward-looking army and navy officers alike.

For all their advantages, the Congreve rockets had drawbacks. The most important was a well-deserved reputation for erratic flight, which some-times made them wildly inaccurate. Part of the accuracy problem was the rocket's center of gravity, which shifted steadily forward as the fuel burned away. Part of it was the shape of the rocket body and the position of the exhaust nozzles, which were seldom perfectly symmetrical. The largest part of the problem, however, was the stick. Like the Indian rockets on which they were based (and virtually all other rockets that came before them), Congreve rockets used a long wooden guide stick to keep them stable in flight. The stick, up to 15 feet long in heavy rockets, made Congreve's weapons cumbersome to handle and vulnerable to air currents while in flight. It also, because it was mounted off-center, tended to throw the rocket off course even when the air was still (see Figure 2.2). Congreve reduced the balance problem in 1815 by mounting the stick in the center of the rocket's base plate and directing the exhaust through a ring of small nozzles around the edge of the plate. Even when centered, however, the stick was never perfectly centered, perfectly stiff, or perfectly straight, and the rockets continued to have a reputation for erratic flight.

William Hale introduced an improved version of Congreve's rocket around 1840. Like Congreve's later designs, it used multiple exhaust vents evenly spaced around the circular base plate. Unlike any previous rocket, however, it used small metal vanes to deflect the exhaust gasses and cause the rocket to spin around its long axis like a rifle bullet. Hale spun his rocket in order to stabilize it: the spinning evened out the effects of not-quite-symmetrical rocket tubes and shifting centers of gravity. Most important, the spinning eliminated the need for a guide stick, which made Hale's rockets more portable, as well as more accurate, than Congreve's.

The British armed forces, though at war in China, Afghanistan, and elsewhere in the 1840s, did not immediately adopt Hale's improved rocket. They clung to the familiar Congreve, as they had clung to the long-serving "Brown Bess" musket, long after newer and better weapons became available. Unable to drum up interest in his native country, Hale sold the manufacturing rights to his rocket to the United States for $20,000—

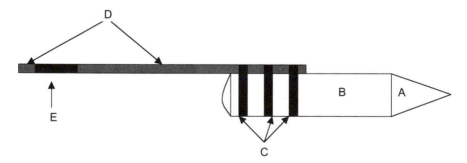

Figure 2.2: A typical early Congreve rocket, showing the attachment of the guide stick. The casing for the warhead (A) and rocket body (B) was made of iron. When the rocket was assembled for use, the stick (D) would be slid through three soft iron bands (C), which were then crimped tightly around it with special pincers. Congreve rockets made for the British army, like the one shown here, used guide sticks that were divided into 4-foot sections for ease of transport, then assembled in the field using soft iron ferrules (E) to join the sections. Drawn by the author.

a substantial sum now, and an immense one then. The first troops to use the Hale rocket in combat were, therefore, the American expeditionary force dispatched to Veracruz in 1847, during the Mexican-American War. Union and Confederate forces both made occasional use of rockets (both Congreve and Hale types) during the American Civil War. The Russian, Italian, Hungarian, and Austrian armies all acquired and used Hale rockets in the 1850s and 1860s, and the British officially adopted them in 1867. Having made the transition, the British military proceeded to cling to the Hale as fiercely as it had to the Congreve. Hale rockets remained in active service until 1899 (well after it, too, had been rendered obsolete) and was not formally stricken from the Royal Army's inventory until 1919.

Britain's long use of Hale rockets was not solely a result of inertia. The wars that Britain fought in the last third of the nineteenth century were small, localized conflicts with native troops in Africa and South Asia. Hale rockets could still be effective against enemies armed with muskets and smoothbore cannon, and they could be carried by pack animals into places that no wheeled gun carriage could reach. On the battlefields of Europe, however, the day of the black powder rocket was essentially over by 1870.

Congreve's rockets had caused a sensation in the first decade of the 1800s because they offered significant advantages over traditional gun artillery. By 1870, however, the situation had been reversed. A series of mid-century breakthroughs in cannon design meant that the best gun artillery had greater range, greater accuracy, and more striking power than the best rocket artillery. Rockets could still be fired faster than cannon, but the gap closed significantly as muzzle-loaded cannon firing balls gave way to breech-loaded cannon firing shells. High-velocity shells even mimicked the high-pitched shrieking noise that made rockets unnerving to the soldiers they were fired at. Rockets played little or no role, therefore, in the turn-of-the-century conflicts that signaled the emergence of modern warfare: the Sino-Japanese War (1894–1895), the Spanish-American War (1899), the Anglo-Boer War (1899–1901), and the Russo-Japanese War (1904–1905). As a weapon, the black powder rocket was dead.

NINETEENTH-CENTURY CIVILIAN ROCKETS

Bombarding enemy troops and fortifications in wartime was only one among many uses to which black powder rockets were put in the nineteenth century. They were also served civilian ends: entertainment, signaling, whaling, and maritime rescue work.

"Skyrockets" continued, of course, to be essential parts of fireworks displays. Indeed, they became steadily more common over the course of the century, as new production techniques lowered their cost and made small versions available to the expanding middle classes as well as to the wealthy and the powerful. Public events sponsored by national governments and large cities—the Great Exhibition of 1851 in London, the Exposition Universelle of 1867 in Paris, and the Columbian Exposition of 1893 in Chicago—continued to be celebrated with elaborate fireworks displays. Especially in the United States, however, modest displays mounted by small towns and individual families also became common, particularly on the Fourth of July.

Signaling rockets were, in a sense, fireworks put to a different use. Carried aboard ships and stored at lighthouses and lightships, they were widely used as emergency signals. Trailing fire as they rose into the sky, then exploding in a burst of red or white light, they could be seen for miles and were virtually impossible to ignore or dismiss. Whether fired just offshore or far out to sea, they served to alert potential rescuers and to guide them to a ship in trouble. The keepers of remote offshore lighthouses, who were cut off from the mainland whenever storms made it impossible to launch a boat, also used signaling rockets to communicate. A white rocket might, by a prearranged code, mean that all was well at the lighthouse, while a red rocket might signal an injury or mechanical failure. The development of radio communication around 1900 diminished the role of signaling rockets but did not eliminate it. The crew of the *Titanic* sent distress calls by radio as the ship foundered on the night of April 14, 1912, but they also fired eight signal rockets in less than an hour.

The use of rockets in whaling was, unlike their use in fireworks, a nineteenth-century innovation. Whaling in the early 1800s was done from small boats, propelled by oars or sails and launched from shore installations or large sailing ships. The whaler's harpoon—typically a spear with a barbed iron head, a slender iron shaft, and a wooden handle—was attached to a long rope that paid out from the boat as the harpoon was thrown at the whale. The purpose of the harpoon was not to kill the whale but to imbed itself in the whale's flesh and hold fast. The rope trailing from the harpoon could then be attached to the boat, which became a powerful drag on the whale. The whalers, now tethered to their quarry, could wait until the whale became exhausted, then pull alongside it and kill it with a long, slender spear called a "lance." Hitting a moving, half-submerged whale from a moving, pitching boat took extraordinary skill, and first-rate harpooners (like the fictional Queequeg and Tashtego in *Moby-Dick*) were highly sought after. Even the best harpooners, however, could not guarantee that a harpoon that hit its target would stay in place. A harpoon thrown with less than full force, or at the wrong angle, could easily glance off the whale's skin or imbed itself so shallowly that it would pull out when the whale tried to swim away. Stories of whales that were "darted," only to be lost, were staples of whaling lore.

The attraction of a rocket-powered harpoon is easy to understand. It would, if it worked correctly, hit harder and penetrate deeper than a traditional hand-thrown model, increasing the chances that it would remain imbedded. William Congreve himself developed one around 1820: a 2-pounder rocket with a center-mounted guide stick extending back from its base plate. Used aboard the whaling ship *Fame* on an 1820–1821 voyage, it played a role in the capture and killing of ten whales. Congreve, along with an artillery officer named Colquhoun, received a joint patent for it in 1821. Details of how Congreve's rocket harpoons were launched are scarce, but the apparatus was presumably the same as that used by the Royal Navy to launch conventional Congreve rockets from small boats: wooden troughs or metal tubes. Later images, depicting a rocket harpoon but not necessarily the Congreve type, show the harpooner balancing a metal launching tube on his shoulder. Thomas Roys, an American whaling ship captain, patented a larger rocket harpoon in 1861 and spent the next several years attempting to refine it. Fired from a deck-mounted gun, it failed to gain a large following among whalers.

The rocket harpoon's brief day in the sun ended in 1864, when Norwegian whaler Sveyn Foyn developed a gun-fired harpoon with an explosive charge behind its barbed head. Foyn's harpoon simultaneously killed the whale and fixed a rope to the carcass—a huge increase in efficiency. It made traditional whaling techniques virtually obsolete and, along with the motor-driven "killer boat" (another Foyn invention), became the technological basis of modern whaling.

The line-carrying rocket, like the rocket harpoon, was an early nineteenth-century innovation. Henry Trengrouse, witness to a December 1807 shipwreck near his home in southwest England, developed one early version shortly afterward. Trengrouse's logic was elegant. Most ships were wrecked within sight of shore, but wind and sea conditions often made it difficult for the crew to reach shore or for rescuers to row to their aid. A line stretched from ship to shore would improve the odds, and a rocket could easily carry one hundreds of yards, even in the teeth of a gale. He argued that the rockets should be fired from ship to shore, so that onshore winds (the cause of most shipwrecks) would aid rather than retard its flight.

Figure 2.3: A Turkish lifesaving crew demonstrates the use of a line-carrying rocket on the shores of the Black Sea in the 1880s. The rocket, already assembled and placed in its launching trough, is visible at the far right. Courtesy of the Library of Congress, Abdul-Hamid II Collection, image number LC-USZ62-82129.

Trengrouse's line-carrying rocket excited little interest among senior officials of the Royal Navy, but it (or the 1817 pamphlet he wrote about it) did catch the attention of William Congreve. The veteran rocket-maker developed his own lifesaving rockets in 1822, adapting 20- and 32-pound military rockets by adding a grappling hook at the nose and an attachment point for a line at the tail.

John Dennett—like Trengrouse, a resident of England's wreck-strewn southern coast—may also have been inspired by the 1817 pamphlet. He tested a series of shore-to-ship lifesaving rockets in 1826–1827 and invited officers of the Royal Army and Navy, as well as the local Coast Guard, to watch. The rockets impressed Dennett's expert witnesses and were soon installed at three Coast Guard stations. When, in 1832, they were instrumental in the rescue of nineteen men from the wreck of the merchant ship *Bainbridge*, Dennett won a national reputation and a government contract to supply more rockets. Twenty years later, in 1853, more than 120 Coast Guard stations around Britain were equipped with them. The Dennett rocket was supplemented, beginning in the late 1860s, by the Boxer rocket. Invented by and named for Colonel E. M. Boxer of the British army, it had two stages and a centrally mounted guide stick. Originally designed as a military rocket, it was withdrawn from service in 1867 because its warhead had an alarming tendency to explode before the rocket left the launching tube. Fitted with an inert "warhead" and a rope, however, the Boxer rocket became a valuable lifesaving tool. Its two-stage design gave it extra range and gentler acceleration, which reduced the chances of the rope breaking.

Both types of shore-to-ship rockets remained in active service for decades (see Figure 2.3). The thirty-six sailors pulled from the wreck of the *Irex* in 1890 rode to the top of a 400-foot cliff along a lifeline put in place by a Dennett rocket. The Boxer rocket was still in use at the beginning of World War II, and was retired only when lighter, more portable systems became available. Most other seafaring nations developed or bought similar systems, and the fiery trails of signaling and lifesaving rockets became a common sight along the world's most dangerous coastlines. Few who watched them would have suspected, however, that a handful of scientists were already exploring the idea of using rockets to reach other worlds.

3

The Birth of Modern Rocketry, 1900–1942

◆

Black powder rockets had, literally and figuratively, gone as far as they could go by the late nineteenth century. The metal-cased, machine-packed, spin-stabilized rockets of the 1890s were a vast improvement on those that had tormented the British in the Mysore Wars a century earlier, but they were also a technological dead end. Black powder generated too little thrust to carry a standard-sized rocket more than a mile or two. Very large powder rockets posed significant engineering problems: how to pack the powder, how to ensure that it burned evenly, and how to prevent the rocket body from flexing while in flight. Increasing the size of rockets also made them more difficult to transport and launch: a serious drawback, since whaling, rescue, and military rockets all had to be portable. Rockets capable of crossing oceans or reaching other worlds (even the relatively nearby moon) were, to all but a few, literally unimaginable. Just *how* unimaginable is evident in the popular culture of the time.

Jules Verne and H. G. Wells were the grandfathers of modern science fiction. Over the half-century between the end of the American Civil War and the beginning of World War I, they produced a steady stream of novels featuring exotic technology and fantastic journeys. Their imaginations were fertile and wide-ranging. *Nautilus*, the submarine in Verne's *Twenty Thousand Leagues under the Sea* (1870), has the look, feel, and performance of a modern nuclear submarine. The airship in *Clipper of the Clouds* (1886) anticipates the dirigibles of the early twentieth century. Wells, in his first "scientific romance," imagined a time machine that would allow his hero to witness the future of human evolution and see the Earth as a half-frozen planet orbiting a dying Sun. Writing just before World War I, Wells anticipated tanks in "The Land Ironclads" (1904), aerial bombing in *The War in the Air* (1908), and nuclear weapons in *The World Set Free* (1914). Both Verne and Wells wrote about imaginary trips to the moon, but neither used rockets to power their heroes' spacecraft. Verne, in *From the Earth to the Moon* (1865) and its sequel *Round the Moon* (1869), sends his heroes to space in a hollow cannon shell fired from a gigantic gun built near Tampa, Florida. Wells, in *The First Men in the Moon* (1901), has his heroes smear the bottom of their spacecraft with "Cavorite"—a (fictitious) substance that blocks gravity the way that rubber blocks electric current. Though gifted with two of the most vivid imaginations of their day, neither Verne nor Wells could conceive of a rocket powerful enough to leave the Earth and reach the moon.

The birth of modern rocketry changed not only how rockets were built but also how they were thought about. It was the result of abandoning black powder, but also of abandoning the idea that rockets *had* to be small, portable devices for carrying modest payloads over short distances. It is hardly surprising that the three men who laid the theoretical foundations of modern rocketry—Konstantin Tsiolkovsky, Robert Goddard, and Hermann Oberth—had soaring imaginations as well as matchless technical skills.

KONSTANTIN TSIOLKOVSKY

Karl Marx, surveying Europe in the middle of the nineteenth century, thought Russia the last place where a socialist revolution was likely to begin. It was, in the last decades of the century, an equally unlikely place for a technological revolution. The glittering eighteenth century, which had begun with the reign of Peter the Great and ended with that of Catherine the Great, was long past. Alaska, the last piece of a once-promising New World empire, had been sold to the United States in 1867. The Industrial Revolution, which had enriched Western Europe for decades, had barely gained a foothold in Russia. A slow unraveling of military and political power, which would end in military humiliation and political revolution soon after 1900, had already begun.

Konstantin Edvardovich Tsiolkovsky was as unlikely a leader for a technological revolution as Russia was a setting for one. Born in 1857, the son of a forester-turned-clerk, he grew up in a small village southwest of Moscow. Motherless and almost totally deaf by the time he was fourteen, he immersed himself in books: physics, astronomy, and mathematics, but also the novels of Jules Verne. He left home at sixteen and spent the next three years in Moscow, renting a corner of someone else's rented room to sleep in and living off cheap brown bread and water. Self-education became Tsiolkovsky's full-time job during the three years he lived in Moscow. He attended scientific lectures, performed chemical experiments, and, above all, read: more physics and astronomy, higher mathematics, chemistry, and philosophy. He left Moscow at nineteen to take a teaching job in the town of Borovsk, and at twenty-five moved on to another teaching job in Kaluga. There he remained—a schoolteacher in a tiny, backwater village a hundred miles from Moscow—until fame caught up with him in his old age. In 1919, when he was sixty-two, the still-new Bolshevik government appointed him to the Socialist Academy (later the Soviet Academy of Sciences) and awarded him a pension generous enough for him to retire from teaching and devote his time to research.

Tsiolkovsky had, by the time the Bolsheviks discovered him, already devised and published his most important ideas. *Free Space* (1883) alluded to orbiting space stations, described the sensation of weightlessness, and outlined the use of rocket engines for propulsion in space. "On The Moon" (1887) described the sensation of walking on the moon and seeing the Earth from a quarter-million miles away. *A Dream of Earth and Sky* (1895) returned in detail to the subject of space stations, portraying them as orbiting utopias whose inhabitants would find freedom from political and social inequity as well as from gravity. All three works were fiction, but, as in Verne's best-known works, slabs of scientific and technical detail often overshadowed the plot and characters. Tsiolkovsky's most important work, however, was a paper titled "Exploring Space with Reactive Devices" that appeared in the *Scientific Review* in 1903. A tour de force of closely reasoned arguments and detailed calculations, it was a how-to manual for a brand-new technology: the high-altitude rocket.

Tsiolkovsky calculated that breaking free of Earth's gravity and reaching orbit would require a velocity of 5 miles per second, or 18,000 miles per hour. Conventional black powder rockets, he concluded, had no hope of achieving that kind of performance. A rocket bound for orbit would have to carry a more potent fuel and (because oxygen was scarce in the upper atmosphere and absent in space) a supply of oxygen to make combustion possible. He proposed hydrogen as a fuel, for its lightness and volatility, and liquefied oxygen as an oxidizer. Carried in tanks inside the rocket's body, they would be pumped into a metal combustion chamber and ignited, producing gasses that would be funneled out the tail of the rocket to produce thrust. Tsiolkovsky's design worked on the same physical principles as black powder rockets, but from an engineering standpoint it was radically different. Powder rockets were bodies filled with fuel, with a space left in which combustion took place. Tsiolkovsky's proposed rocket was a body filled with mechanical components: tanks for the liquid fuel and oxidizer, pumps and piping to move them, a separate combustion chamber in which they combined, and an igniter to set them burning. It was a technological *system*: a group of separate, but closely integrated, components designed to work together. The advantage of such a design, Tsiolkovsky realized, was that any individual component could be modified (up to a point) independently of the others. Substituting a more potent fuel, a more powerful pump, or a differently shaped combustion chamber or exhaust nozzle could allow designers to improve performance without starting from scratch.

Tsiolkovsky also argued, in "Exploring Space," for the importance of multistage rockets. He realized that the key to colonizing space—his ultimate goal—was to design a rocket that could accelerate its own weight *and* a useful payload to 5 miles per second or more. A huge single-stage rocket would have to lift its own enormous weight all the way to orbit. Even when only 10 percent of its propellant remained, for example, it would still have to lift the weight of tanks designed to hold ten times that much propellant. Using multiple stages meant that the thrust (and weight) of the rocket was divided into discrete packages. The first stage would accelerate itself, the subsequent stages, and the payload to a given speed and lift them to a given altitude. Its propellant exhausted, its pumps and combustion chamber reduced to dead weight, it would then be discarded. The second stage would then take over, carrying a much smaller load closer to orbit. Tsiolkovsky's multistage design was not only efficient but also flexible. Adding stages or replacing less powerful ones with more powerful ones would, like upgrading individual components, give substantial improvements in performance.

Even in his later years, when he worked with the full backing of his government, Tsiolkovsky never built or flew a rocket to test his ideas. He was a theorist, not an experimenter or an engineer. Soon after his "discovery" by the Bolsheviks, however, others who *were* engineers would begin building and flying rockets shaped by his ideas. The designers of those rockets were almost all Soviet citizens. Tsiolkovsky was a prolific writer, and continued to expand "Exploring Space" well into the 1920s, but his work was virtually unknown outside the USSR until the 1930s. Meanwhile, in the United States and Europe, others were working along similar lines.

ROBERT GODDARD AND HERMANN OBERTH

Robert Hutchings Goddard and Hermann Oberth shared Tsiolkovsky's boyhood love of the works of Jules Verne, his fascination with rockets, and his near-obsession with the idea of giving humans access to outer space. Beyond that, however, they had little in common with the self-educated schoolteacher from Kaluga. Goddard and Oberth were born a generation later than Tsiolkovsky, in 1882 and 1893 respectively. More to the point, they were born and raised in the two most technologically sophisticated countries of the late nineteenth and early twentieth centuries: Goddard in the United States, and Oberth in Germany. Goddard earned a bachelor's degree from Worcester Polytechnic Institute in 1908, and a master's and doctorate in physics from Clark University (also in Worcester, Massachusetts) in 1910–1911. He taught at Princeton for three years before returning to Worcester in 1914 to join the physics department at Clark. Oberth, urged by his father to study medicine, pursued physics and astronomy instead, first at the University of Munich, and later at Göttingen and Heidelberg.

Goddard published his first major work in 1919. A sixty-nine-page treatise titled "On a Method of Reaching Extreme Altitudes." It was a serious technical study of how two-stage, solid-propellant rockets could be used to lift scientific instruments high into the Earth's atmosphere, and ended with brief discussions of liquid-propellant rockets and the possibility of sending a rocket to the moon. Funded and published by the Smithsonian Institution, Goddard's pamphlet-sized work oozed respectability. The argument was dense, the writing dry, and the pages studded with equations and tables of data. It was, in other words, a model of respectable scientific writing, and it addressed an important scientific problem: how to gather atmospheric data from altitudes higher than the 7 miles balloons could reach. It was also, for anyone other than a physicist or would-be rocket builder, staggeringly dull stuff.

The appendix dealing with the moon rocket—three pages out of sixty-nine—was something else: shorter, less technical, and easy to sum up in a few words. When a Smithsonian news release briefly mentioned the idea on January 11, 1920, journalists seized on it. Goddard became an overnight celebrity, mentioned in dozens of breathless headlines and dubbed "the moon-rocket man" or a "modern Jules Verne." The *New York Times*, after running a thorough description of Goddard's pamphlet on January 12, chided him on its January 13 editorial page for making what the editors saw as an elementary technical error. It was obvious, the *Times* scolded him, that a rocket could not work in outer space: there was nothing for the rocket exhaust to push against.[1] Goddard, quiet and serious to the point of stiffness, was horrified by the attention and did everything possible to avoid it.

Oberth's first major work, the doctoral thesis he wrote at Heidelberg, also triggered a personal catastrophe of sorts. The thesis was a detailed theoretical demonstration that space travel was possible, coupled with an analysis of how space travel was likely to affect the human body. It was an audacious piece of work, but it straddled two academic disciplines that did not yet exist: astronautical engineering and space medicine. It fit into no existing department, and there was no single faculty member at Heidelberg who had the expertise to pass judgment on the entire work. Organizational rigidity—present in all universities, but especially in German ones—magnified the problem. The Heidelberg faculty rejected his thesis in 1922, denying him the aura of academic respectability that being "Herr Doktor Oberth" would have given him.

Goddard and Oberth were men with similar minds but very different personalities. Their reactions to the less than gratifying receptions of their work reflect those differences. Goddard became guarded and secretive. He began to build and test rocket motors and, eventually, entire rockets, but he "published" the results of his work only in confidential reports to organizations (notably the Smithsonian) that funded him. He had assistants—friends, family members, and employees—but not collaborators. Convinced (as the Wright Brothers had been) that others would steal his ideas and profit from them, he rebuffed fellow rocketeers who inquired about his work.

Oberth, in contrast, became a tireless and skilled self-promoter. Denied a doctorate, he told the Heidelberg faculty that he would "become a greater scientist than some of you, even without the title of doctor" (Crouch 1999, 36). Using borrowed money, he published his thesis in 1923 as a slender ninety-two-page book titled *Die Rakete zu den Planetenraum* ("The Rocket into Interplanetary Space"). It sold well enough to justify a second printing in 1925, and was followed by a second book, *Wege zur Raumschiffart* ("Ways to Space Flight"), in 1929. He formed informal contacts with science writers like Max Valier and Willy Ley, who used his ideas as the basis for popular, nontechnical works, and served as chief technical consultant on *Frau im Mond* ("The Woman in the Moon")—a 1929 science fiction film by noted director Fritz Lang.

---

1. The editors of the *Times* were wrong. A rocket moves, as Goddard well knew, because the escaping exhaust gas pushes against the rocket, not the ground beneath it or the air behind it. The *Times* never published a formal retraction, but acknowledged in 1969 (after the first moon landing) that it "regret[ted] the error."

Goddard, despite his professional isolation, achieved great success. He had access to laboratories, machine shops, skilled assistants, and (thanks to his skill as a fund-raiser) more money than all but a handful of American scientists. Over the course of the 1920s, he built a series of rocket motors and ran "static tests" in which a robust frame held them in place and instruments measured the thrust they produced. He began with solid-propellant motors, which he had described in detail in his 1919 pamphlet. After a few years he moved on to liquid-propellant motors, which in 1919 he had mentioned only in passing. His first success came in December 1925, when a small motor burning gasoline and liquid oxygen ran for 27 seconds on a test stand in a Clark University laboratory. The motor had, Goddard noted with satisfaction, produced 12 pounds of thrust: enough to lift its own weight.

The next step was to build a flyable rocket around the motor, and Goddard spent the next three months doing just that. The result was a fragile, ungainly, purely functional machine: two cylindrical units, one well above the other, joined by a pair of slender pipes (see Figure 3.1). The lower unit, capped by an asbestos-covered cone to protect it from the hot exhaust, contained two tanks: one each for gasoline and liquid oxygen. The upper unit contained the igniter, combustion chamber (a steel tube lined with an aluminum oxide compound), and exhaust nozzle. The pipes carried oxygen and gasoline from the tanks to the chamber, and gave the rocket structure. There was no sheet metal body to enclose and streamline the mechanical components; Goddard had calculated that, had he installed one, the rocket could not have lifted it. As it was, the rocket could barely lift itself. Test flown on March 16, 1926, at a farm belonging to one of Goddard's cousins, it flew for 2.5 seconds, reaching a maximum altitude of 41 feet and traveling 184 feet from the launch point. Modest as it was, it was the first flight of a liquid-propellant rocket: proof that the basic concept was sound and capable of being improved upon.

Goddard threw himself into making such improvements. He built a series of increasingly sophisticated rockets, flying them—as he had flown the first one in 1926—from the edge of the cabbage patch at his cousin's farm. The last of these launches, in July 1929, reached an altitude of more than 80 feet. It was enough to alarm the neighbors, attract the attention of the local police and press, and convince Goddard that it was time to find a more isolated test site. The following summer, Goddard moved his operation to a rented house and 8 acres of land just outside Roswell, New Mexico (see Figure 3.2). His wife, four assistants, and a boxcar-load of equipment went with him. Money from the Carnegie Institute and philanthropist Charles Guggenheim funded the move and a four-year program of research and test flights. Happily isolated from anyone who was remotely interested in rockets, Goddard picked up the pace of his research. By 1935, he was flying

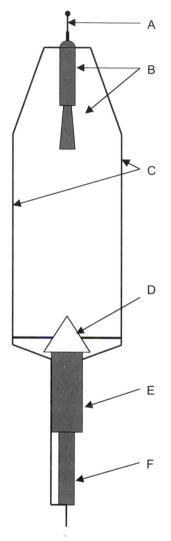

Figure 3.1: A schematic diagram of Robert H. Goddard's 1926 rocket, the first in history to use liquid propellants. The igniter (A) and cobustion chamber with exhaust nozzle (B) form the upper section of the rocket. The oxygen (E) and gasoline (F) supplies, capped with a protective asbestos cone (D), form the lower section. A framework of metal pipes (C) links the two and carries propellants to the combustion chamber. Drawn by the author.

"A-Series" rockets that burned gasoline and liquid oxygen as propellants and reached altitudes of up to 7,500 feet. They were the first rockets in history to be stabilized by onboard gyroscopes, the first to use heat-proof carbon vanes to steer the rocket by deflecting its exhaust stream, and the first

Figure 3.2: Robert H. Goddard tests a liquid-propellant rocket at his isolated research station outside Roswell, New Mexico, in the 1930s. The launch pad and
support structure for the rocket are visible on the left, at the end of the dirt road.
Goddard stands in the door of his "control room," observing through a telescope
with his left hand on the launch controls. Courtesy of NASA Headquarters, image
number 74-H-1245.

to employ a variety of other innovations. Goddard continued, however, to
keep his breakthroughs a closely held secret. Private notebooks, patent applications, and confidential reports to his backers remained his favorite
places to record his work. The only significant exception to this pattern
was *Liquid Propellant Rocket Development,* a 200-page distillation of what he
had learned in three years at Auburn and six years at Roswell. Published by
the Smithsonian in 1936, it offered anyone who could read and understand
it a graduate-level education in rocket engineering. It was his second, and
last, major publication in the field.

Hermann Oberth's rocket-building career was shorter and less impressive. It began and ended with a single machine, which he agreed to build
and launch as part of a publicity campaign for the premiere of *Frau im
Mond.* The spacecraft depicted on-screen in Lang's film was a group of in-

tricately detailed stage sets based on the "Model E" rocket described in Oberth's 1923 book. The rocket that he intended to launch was far smaller and far less powerful, but it would have been the first liquid-propellant rocket flown in Europe. Oberth was soon forced, however, to confront the fact that he had no idea how to translate his theoretical work into a functional rocket. Goddard's modest 1926 rocket worked because an ingenious array of valves, floats, levers, and wires regulated the flow of gasoline and oxygen into the combustion chamber. It was the work of a man who had access to well-equipped machine shops and experienced machinists, and who was himself familiar with the business of "bending metal." Oberth, setting out to build a more ambitious rocket virtually from scratch, had neither Goddard's experience nor his access to resources. He and two hired assistants worked for four intense months, but produced only explosions and a single successful static test of a small prototype motor.

Oberth publicly failed where Goddard had privately succeeded, but his immediate impact was far greater. Germany's growing community of amateur rocketeers cherished the mixture of hard technical details and bold vision in Oberth's books, and they admired him as a symbol of Germany's continued leadership in scientific and technical fields. Germany struggled, in the 1920s, with a fragile new system of government and economy-crushing payments imposed on it by the Allied powers after World War I. It had been stripped of its overseas empire and most of its armed forces, and forbidden by the terms of the peace treaty to rebuild either one. Rocketry became a source of national pride at a time when such pride was in short supply, and Oberth (according to the works of Valier and others) was the world's foremost expert on rockets. Whether he, personally, could build the machines he wrote about was beside the point. It was, for the members of Germany's amateur rocket societies, enough that he existed and that he was German.

THE ROCKET SOCIETIES

The British have a word for those whose leisure time revolves almost exclusively around a particular subject: "enthusiasts." The closest equivalent in the American vocabulary is "buffs," but except in isolated instances ("railroad buffs" and "Civil War buffs," for example) it does not carry the same connotation of total, all-consuming interest. The members of the amateur rocket societies that formed in the 1920s and 1930s deserve to be called "enthusiasts." They lived for the hard, dirty, frequently dangerous work of building and flying rockets and dreamed of a day when rockets would open the road to worlds beyond Earth. Inspired by the work of Tsiolkovsky, Goddard, and Oberth, the members of the various rocket societies built and flew dozens of rockets in the late 1920s and 1930s. The societies became a testing ground for new technologies and a training ground for the designers who would dominate rocketry in the 1940s, 1950s, and 1960s.

The first, and most influential, of the rocket societies was the Verein für Raumschiffahrt ("Society for Space Travel"), headquartered in Berlin and often known simply as the VfR. Founded by ten people meeting in a Breslau bar in July 1929, it had grown to 500 members within a year and 900 within two years. Oberth was a member, as was Rudolf Nebel—one of his assistants on the failed *Frau im Mond* rocket. Robert Esnault-Pelterie, a Frenchman who wrote the first serious study of long-range ballistic missiles, joined as well. Willy Ley and Max Valier, who had helped to publicize Oberth's ideas, were among the founders. The VfR had two goals: raising public awareness of space travel, and advancing the state of the art in rocketry. In pursuit of the first, they published a widely read newsletter, *Die Rakete* ("The Rocket"), and by 1930 were organizing rocket exhibitions. In pursuit of the second, they took over an abandoned (and rent-free) army post on the outskirts of Berlin: 300 acres of open space for test flights, along with buildings for workshops and (for some members) living space. With one eye on the future that they hoped to create, they named it the *Raketenflugplatz* ("Spaceport").

The VfR's first important launches took place at the *Raketenflugplatz* in May 1930. A rocket powered by gasoline and liquid oxygen made two flights within three days, reaching nearly 60 feet on the first and close to 200 feet on the second. Two years, 270 static tests, and 87 flights later, rockets launched by the VfR had reached altitudes of a mile and covered horizontal distances of three miles. Valier was dead by then, killed by a shard of flying metal when a rocket-powered car he was working on exploded in 1930, but new members had continued to join. The most important of them, in retrospect, was a young aristocrat with the broad shoulders of an athlete and the face of a movie star. His name was Wernher von Braun, and he was seventeen when he signed on with the society in 1929. Von Braun's role in the VfR was relatively modest—his studies at the Berlin Institute of Technology took up much of his time—but it gave him practical experience and put him in contact with the elite of the German rocket-building community. The VfR, always short of money, went bankrupt in 1934, but by then von Braun had moved on to graduate studies and a new relationship with the Army Ordnance Department.

Rocket societies also emerged in the Soviet Union in the 1920s, inspired by the writings of Tsiolkovsky as the VfR had been inspired by those of Oberth. The most important of them merged, between 1929 and

1931, into the Group for the Study of Reaction Motors (GIRD), which had major branches in Moscow and Leningrad and minor ones throughout the western Soviet Union. The Moscow branch, known as MosGIRD, built and tested the first liquid-propellant rocket motors developed in the USSR. It achieved its first successful launch in August 1933, sending a rocket simply named the "09" on an 18-second flight to 1,300 feet. A little more than three months later, in November 1933, a more powerful rocket named the GIRD-X reached an altitude of 3 miles on *its* inaugural flight.

The sheer talent represented by MosGIRD, especially, was staggering. It included Valentin Glushko and Mikhail Tikhonravov, both of whom became major figures in Soviet rocketry. The man behind the GIRD-X was Fridrikh Tsander, whose innovations included a system to cool the combustion chamber by circulating the propellants around it in pipes. The most significant member of MosGIRD turned out, however, to be the designer of the modest 09—a young engineer named Sergei Korolev. Korolev, Glushko, and the other leaders of MosGIRD, along with the leading lights of the Leningrad branch of GIRD (LenGIRD) and the government's Gas Dynamics Laboratory, were folded into a new organization in 1933. Called the Reaction Propulsion Research Institute (RNII), it was under the direct control of General Mikhail Tukhachevsky, the head of the Red Army's ordnance department, and was designed to advance the development of all kinds of military missiles.

The American equivalent of the GIRD and the VfR was founded in 1931 as the American Interplanetary Society. It began as a group of science-fiction fans with dreams of space travel, and its members contented themselves, at first, with talking about space travel and publishing a mimeographed newsletter. Attempts to build working rockets came later, after founding member G. Edward Pendray visited Berlin and watched Rudolf Nebel run a static test on a small rocket motor. Determined to build their own rocket, Pendray enlisted ex-Navy machinist Hugh Pierce, scrounged parts, and improvised a launch site on Staten Island. The society's first rocket lifted off in May 1933, reached an altitude of 250 feet, then tumbled into the shallows of New York Bay after an exploding oxygen tank caused the engine to fail. Undaunted, Pendray and his colleagues regrouped. The following year, 1934, the society renamed itself the American Rocket Society (ARS), transformed the cheaply printed newsletter into a magazine, and began work on more, better rockets. It also acquired two new members who would become the creative forces behind those rockets: John Shesta and James Wyld. Shesta was a university-trained engineer—a graduate of Columbia—when he joined in 1934. Wyld was still a senior at Princeton, tinkering with rockets in his spare time, when he came aboard in 1935.

It was a rocket motor designed by Wyld, first tested in December 1938 and improved in 1941, that brought the society real attention. Weighing only 2 pounds, the motor was small enough to hold in one hand or tuck in a briefcase, but it produced an astonishing 90 pounds of thrust. Told that the U.S. government might be interested in such a motor, but that it did business only with corporations, members of the society quickly formed one. Shesta, Wyld, Hugh Pierce, and Lovell Lawrence (who had made the initial contact with the government) thus became the entire staff (managers as well as employees) of Reaction Motors, Incorporated.

THE REVIVAL OF MILITARY ROCKETRY

Serious military interest in rockets had faded after the American Civil War, and remained dormant well into the twentieth century. Only the French had used them during World War I. Interest gradually revived in the mid-1930s, however, as military leaders gradually became aware of the work of individuals like Goddard and groups like the VfR and MosGIRD. The American Rocket Society's decision to form Reaction Motors, Inc., and peddle its lightweight rockets to the U.S. armed forces was atypical, in that it was the rocketeers who took the initiative. In Germany, in the Soviet Union, and on the west coast of the United States, it was the military that sought out the rocketeers.

Military interest in rockets was not widespread. It began, in all three countries, with individual officers who saw potential in the liquid–fuel rockets of the late 1920s and early 1930s. In the USSR and in Germany, the key figures were former artillery officers turned ordnance experts: General Tukhachevsky and Colonel Karl Emil Becker. Both officers came from branches of their armies that respected scientific and technical expertise. Both envisioned rockets, not surprisingly, as a form of artillery. In the United States, however, the story unfolded differently. Military interest in rockets focused on their potential as a supplementary power source that would help heavily laden aircraft get off the ground quickly. Commander (later Captain) Robert Truax, head of the Navy's rocket program, had short aircraft–carrier decks in mind. General Henry H. Arnold, commander of what was then the Army Air Corps, planned to enhance the giant, long–range bombers that he believed represented the future of air power.

Government—specifically, military—support gave the rocket designers of the 1930s access to materials, equipment, facilities, and above all money. Arnold began the U.S. Army's rocket research program by authorizing, in mid–1938, a $10,000 grant to a team led by Professor Theodore von Karman and graduate student Frank Malina of the California Institute of Technology (also known as CalTech). It enabled them to turn a makeshift test site on the outskirts of Los Angeles into a rough-but-functional rocket research center: the beginnings of what is now Jet Propulsion Laboratories. The VfR quietly went bankrupt in 1934, but Wernher von Braun and other members who went to work for Becker as civilian employees of the German army were well-supplied and well-funded. Von Braun's team was relocated, in 1937, to a state-of-the-art research and test facility on the island of Peenemünde, on the North Sea, and by 1939 the German government had poured $90 million into the site. The 1919 Treaty of Versailles systematically deprived Germany of tanks, submarines, heavy artillery, and other offensive weapons, but it placed no limits on rockets. Even before the Nazi government began to openly defy the treaty, therefore, Becker's team (eventually taken over by his former assistant, Walter Dornberger) could work on long-range guided missiles perfectly legally.

Using government resources, however, meant accepting government control, and the consequences of doing so could be severe. Von Braun and his colleagues, still driven by their dream of sending rockets to other worlds, could keep working only if they built rockets that could hit targets on Earth. Goddard, who went to work for Truax and the U.S. Navy after Pearl Harbor, put aside his lifelong distaste for collaborating and sharing knowledge for the duration of the war. It was Korolev, however, who paid the heaviest price for the government support he had enjoyed. Soviet leader Josef Stalin began, in 1937, to purge the Soviet state of anyone even remotely suspected of disloyalty. The purges eventually claimed millions of victims, but Tukhachevsky—who had founded the RNII and acted as Korolev's patron—was among the first to be arrested, tried, convicted, and shot. Korolev, guilty by association and denounced by his professional rival Glushko, was sent to a gulag in 1938 and remained there until the end of World War II. Only the intervention of Andrei Tupelov, a leading aircraft engineer and fellow victim of Stalin's paranoia, saved him from hard labor in the mines and the death by exhaustion that would have followed.

The resources of three major military powers, funneled through talents of designers like Korolev, Glushko, von Braun, Malina, and Wyld, produced substantial advances in rocketry in the late 1930s and early 1940s. The breakthroughs made during those years established liquid-propellant and large solid-propellant rockets as viable technologies, and set the stage for their rapid development (for both military and quasi-civilian uses) after World War II.

At CalTech, chemist John Parsons discovered that a stiffly viscous mixture of asphalt and potassium perchlorate made an excellent rocket propellant and—unlike the granular propellants then in use—would not develop cracks if stored in the rocket for long periods of time. Martin Summerfield, another member of the von Karman team, developed a new type of liquid propellant in consultation with members of Robert Truax's team in Annapolis, Maryland. Rather than liquid oxygen, which had to be handled with specialized equipment and began evaporating as soon as it was pumped into a rocket's tanks, Summerfield used fuming nitric acid (standard nitric acid plus nitrogen dioxide) as an oxidizer. Fuming nitric acid produced poor performance when used with gasoline or kerosene, but (as one of Truax's chemists suggested) worked well with aniline. The fuming-nitric-acid/aniline combination also offered a bonus: mixing the two chemicals caused them to ignite spontaneously, eliminating the need for a separate igniter.

Challenged by the Army Air Corps to develop strap-on rocket boosters for aircraft, the CalTech group tested the first ones in 1941, using a lightweight private plane called an Ercoupe. Flown by a volunteer Army pilot, it streaked off the runway and into the sky, demonstrating the value of the new solid-propellant motors. The CalTech group, like the American Rocket Society before them, formed a corporation in order to market their innovations to the government. They called it Aerojet for the same reason that the booster motors they designed were called *Jet*-Assisted Take-Off (JATO) units: to most Americans, the word "rocket" still suggested Buck Rogers comic strips, not serious technology. That attitude changed by the end of World War II, however, and Aerojet became (along with Reaction Motors) one of the nation's two principal builders of rocket engines.

The same process unfolded, with different results, on the other side of the Atlantic. Korolev and his fellow engineer-prisoners designed small solid-propellant rockets for the Red Army and Air Force, and sketched designs for longer range liquid fuel missiles. Von Braun's team made a series of breakthroughs that would, in 1943, make possible the missile they called the A-4 and their Nazi backers dubbed the V-2. Walter Thiel, leader of the team that designed the rocket motor, implemented "film cooling": keeping the throat of the motor cool by letting a thin film of alcohol flow over its inside surface. Von Braun worked with pump manufacturers to develop a lightweight, high-pressure, highly reliable pump to force fuel and oxidizer into the combustion chamber. Other engineers struggled to produce a guidance system that would keep the missile on course, as well as stable in flight. Plans for larger, two- and three-stage missiles with ranges measured in thousands of miles began to take shape on the drawing boards at Peenemünde.

The breakthroughs made in the mid- to late 1930s and the early 1940s led, in time, to rockets that could carry warheads across oceans, instruments into the upper fringes of the atmosphere, and even humans into space. The payoff, however, would not come until the late 1940s and early 1950s. The intervening years would be dominated by World War II, which proved to be a watershed in the development of rocket technology. On one hand, wartime demands for innovative weapons produced technological break-throughs like the V-1 cruise missile, the V-2 ballistic missile, and the first "smart" weapons. On the other hand, most of the millions of rockets fired during World War II were short-range, unguided projectiles. The battlefield rockets of World War II belonged, in this respect, as much to the 1840s as to the 1940s. They were weapons like those developed by Congreve and Hale, raised by a century and a half of improvement to a murderous new level of efficiency.

4

Rockets in World War II, 1939–1945

Rockets were used for centuries on the world's battlefields, principally as a form of artillery. Deployed and fired in concentrated masses, they made up for limited accuracy with their considerable striking power and their incomparable psychological effect on their victims. Eclipsed by the rapid improvement of artillery during the nineteenth century, they made a widespread comeback in World War II. One reason for rockets' renewed popularity was the technological advances made in the 1920s and 1930s: liquid fuels, gyroscopic guidance systems, and so on. Another reason was the development of vehicles—airplanes, trucks, landing craft—whose structure enabled them to carry rocket launchers but not heavy cannon.

Rockets were used in three distinct roles during World War II. The first role was centuries old: barrage rockets, fired rapidly and in quantity in order to saturate a large area of the battlefield in a short time. The second was relatively new, having been pioneered by the French air force in World War I: direct-fire rockets, aimed singly or in small numbers at specific targets. The third role was entirely new: rocket propulsion systems for aircraft and guided missiles. Rockets made a significant contribution to the war in their barrage and direct-fire roles, altering the course of battles and the tempo of entire campaigns. Rocket propulsion had a far smaller impact during the war, but an enormous impact afterward. An extraordinary range of innovative weapons were developed during World War II, but none (even the nuclear bomb) has changed warfare more than the rocket-propelled guided missile.

BARRAGE ROCKETS

Cannon had been the backbone of the world's artillery units for 500 years before World War II. They continued in that role throughout the war itself. When handled by skilled crews, cannon remained unmatched in their ability to deliver accurate, sustained, heavy fire from a distance. Cannon also, however, retained their traditional shortcomings. They were complex and expensive to manufacture, difficult to move on short notice, and (because of their ferocious recoil) capable of being fired only from a solid foundation.

Rocket launchers were less technologically sophisticated than even the simplest cannon. Most consisted of little more than a set of launching rails or tubes, mounted in parallel on a metal frame that could be rotated or tilted in order to aim them. Most of the rockets they fired were equally straightforward: unguided, solid-propellant weapons with diameters under 6 inches and warheads measured in tens of pounds. The individual barrage rockets fired in World War II used more potent propellants and explosives, and more sophisticated fuses, than the barrage rockets of the nineteenth century. Barrage rockets as a *system*, wever, were still nearly as simple as the system developed by Congreve in the early 1800s.

Barrage rockets' simplicity made them an ideal battlefield complement to large cannon. Because they were not precision machines, rocket launchers could be built quickly and cheaply in virtually any well-equipped factory. Because they were relatively light and produced no recoil, they could be mounted on any vehicle larger than a motorcycle. The ease of building and deploying rocket launchers encouraged commanders on both sides of World War II to bombard enemy positions with rockets as a prelude to attack. Electric ignition systems, standard by the 1940s, facilitated such barrages by allowing the rockets from a single launcher to be "ripple fired"—launched one after another at precise split-second intervals. Ripple firing multiplied the psychological impact of rocket barrages, subjecting the target to a steady cascade of explosions.

Germany began developing rocket artillery in the 1930s, as part of the rearmament program begun by the Nazis. The standard German army rocket launcher, first deployed in 1940, consisted of six short, wide tubes arranged in a circular cluster (like chambers in the cylinder of a revolver) and mounted on a lightweight gun carriage. The launcher looked like a stubby six-barreled cannon, and with good reason: it was adapted from a mortar designed to lob smoke and gas shells onto enemy positions. Its name—*Nebelwerfer* (smoke thrower)—was a legacy of that early stage in its development, and was retained as a way of masking the weapon's true function. The *Nebelwerfer* was far from an ideal weapon: its range was limited, its accuracy was atrocious, and the 300-yard smoke trails of its rockets instantly revealed its position for enemy gunners. Like the military rockets of earlier centuries, however, its projectiles took a psychological toll as well as a physical one. Rifle and machine gun bullets, moving at supersonic speed, were invisible, but the *Nebelwerfer*'s rockets arced toward their targets whistling and trailing smoke. Soldiers under attack by them could only take cover and wait for impact, knowing that if they survived they'd have to do it all again moments later. Even those who were not physically injured suffered intense emotional stress.

The *Nebelwerfer*'s capacity for physical destruction was also impressive. The original six-tube model could launch six 150 mm rockets, each with a 5.5-pound warhead, in under ten seconds. The later five-tube model, which fired 210 mm rockets with 22-pound warheads, could hit even harder (although even less accurately). A battery of well-concealed, well-positioned *Nebelwerfers* could saturate a large area with high explosive in a matter of seconds. Used against soldiers massed for an attack, they could be deadly, as Allied troops discovered after the invasion of Normandy in 1944.

The Soviet Union's prewar involvement in rocket research and its preference for simple, robust, mass-produced weapons made it, too, a natural setting for the development of barrage rockets. The Soviet army was the first to deploy a vehicle-mounted multiple-rocket launcher, a weapon that Soviet troops called the *Katyusha* (roughly, "Little Katie") and their German adversaries called the "Stalin Organ." The *Katyusha* consisted of eight parallel steel rails roughly 18 feet long, mounted atop a steel frame that lifted them above the vehicle and held them at the desired launch angle (usually about 30 degrees above horizontal). Each rail carried two rockets: one attached to its top edge and one to its bottom edge. Each rocket, a little over 6 feet long and 5 inches (132 mm) in diameter, could carry a 44-pound warhead about 5 miles. The rockets were inaccurate but, especially when fired in massive quantities at the beginning of an attack, highly effective at breaking up German defenses. Designed in 1938–1939 and tested in December 1939, they were first used in combat during the German invasion of the Soviet Union in the summer of 1941 and remained in active service throughout the war. *Katyushas* could be mounted on tanks or other tracked vehicles, but they were most often mounted on ordinary military trucks—a cheap, durable, readily available platform.

The U.S. Army experimented along similar lines, producing a variety of vehicle-mounted launchers. The first to enter service was the T27 Xylophone, named for the side-by-side arrangement of its eight launching tubes. Variations on the theme included the T27-E2 (a twenty-four-tube successor to Xylophone), the T44 (a 120-tube launcher fitted to amphibious trucks like the DUKW), and the T45 (a fourteen-tube launcher for mounting on jeeps). The most innovative launcher in the U.S. Army inventory was the T34 Calliope: a sixty-tube launcher mounted, in a wooden frame, on the turret of a Sherman tank. Calliope had two significant advantages over truck-mounted systems. First, because the launcher turned with the turret and raised or lowered with the tank's main gun, it could be aimed quickly and easily. Second, compared to trucks and jeeps, tanks were better equipped to withstand enemy counterattacks and fight on their own once their rockets had been fired. Calliope-equipped Shermans were, in theory, capable of jettisoning their launchers in a matter of moments and becoming ordinary tanks again. Until the last months of the war, all U.S. Army rocket launchers fired the standard M8 4.5-inch rocket: short-ranged and highly inaccurate, but effective as a barrage weapon.

The Army's attitude toward multiple-rocket launchers was ambivalent at best. On one hand, the launchers were deployed in both the European and Pacific theaters, and at least one complete artillery battalion was equipped with them. They were used in combat from June 1944 onward, but nearly all multiple-rocket launchers carried official designations beginning with T (for "test")—a sign that they were regarded only as a temporary experiment.

The U.S. Navy and Marine Corps, by contrast, embraced rocket artillery and made extensive use of it. The Marines saw lightweight, vehicle-mounted rocket launchers as artillery support that could be brought into action quickly when assaulting enemy-held beaches. Their training school for rocketeers, established on the Hawaiian island of Oahu early in 1944, graduated its first class in April of that year. The first of six "provisional rocket detachments" was formed the same week. Each detachment consisted of one officer, fifty-seven enlisted men, and (initially) a dozen 1-ton trucks with 1-ton trailers. All six rocket detachments eventually saw action in the Pacific, first at the invasion of Saipan in June 1944 and later in the invasion of the Philippines in late 1944 and the invasions of Iwo Jima and Okinawa in 1945. The Marines developed their rocket tactics through trial and error, learning from battlefield experience how to use rocket artillery most effectively. The most critical lessons involved the vulnerability of the launchers and the unarmored trucks that carried them. The Marines originally deployed their launchers ahead of the front line of troops to maximize range, but soon shifted them back to protect the rocketeers from being overrun by the enemy. They also learned, as Army rocketeers in Europe had learned, to move their launchers immediately after firing in order to avoid "counter-battery fire" by enemy artillery and mortars. Photographs show that the Marines experimented with tank-mounted launchers (offering both protection and mobility), but there is no official record of such a program.

The U.S. Navy's commitment to barrage rockets was even stronger. Indeed, the United States led the world in developing rockets as a naval bombardment weapon. Rockets' relatively light weight and minimal recoil enabled the Navy to mount them on landing craft originally designed to ferry troops onto enemy-held beaches (see Figure 4.1). Rocket-firing landing craft filled a crucial role in amphibious invasions. Designed to operate in shallow water, they could accompany the invasion force to the beach and blanket it with high explosives just moments before the first troops went ashore. Conventional naval bombardment—cannon fire from battleships, cruisers, and destroyers stationed offshore—had to be halted or moved inland when the invasion force neared the beach, for fear of hitting friendly troops. Rocket barrages fired from incoming landing craft could hit the beach itself moments before the assault troops. Enemy troops would thus be forced to remain under cover longer, making it more difficult for them to mount an organized, effective defense.

The Navy first used rocket barrages during Operation Torch—the invasion of North Africa—in 1942, and they soon become a standard part of amphibious operations. They were used extensively in the invasions of Normandy and southern France in 1944, and in virtually every Pacific theater invasion from January 1944 on. The vessels used ranged from Landing Craft Infantry (Rocket) carrying launchers for sixty 5-inch rockets up to Landing Ships Medium (Rocket) carrying launchers for nearly 500. The landings at Iwo Jima in February 1945 were preceded by two complete barrages fired by a line of twelve LSM(R)s. The destructive power of such a bombardment was staggering: more than 10,000 rockets poured onto the beach in a matter of minutes. The psychological effect was equally so: thousands of screaming projectiles trailing fire and smoke as they arced across the sky.

DIRECT-FIRE ROCKETS

Direct-fire rockets were used, in combat, as though they were cannon: the operator pointed the launcher at a target and pulled the trigger. At short ranges, against stationary targets or vehicles with limited mobility, they were accurate enough to be effective. Their light weight and nonexistent recoil meant that they could be carried by individual soldiers or mounted (six or

Figure 4.1: U.S. Navy landing ships converted to carry rocket launchers bombard Japanese positions in preparation for the 1945 invasion of Okinawa. A single ship of this type could discharge as many as 500 rockets in a matter of minutes. U.S. Navy photograph. Courtesy of the Library of Congress, image number LC-USZ62-92435.

eight at a time) on aircraft. A rocket could hit harder, however, than the shells from any gun that a man or a typical airplane could carry. The combination of light weight and devastating power gave direct-fire rockets their appeal. They put the power of a small cannon in the hands of individual soldiers, and enabled fighter planes to destroy targets that would once have demanded a squadron of bombers.

The most famous direct-fire rocket launcher of the war was the U.S. M1A1 type, universally known to American troops as the "bazooka" because of its resemblance to a trombone-like folk musical instrument with the same name. The bazooka was invented in 1942 by Captain (later Colonel) Leslie Skinner, who saw it as a way for infantry soldiers to defend themselves against enemy tanks without relying on artillery support. The bazooka consisted of a steel tube—4 feet long with a 2.36-inch inside diameter—with wooden handgrips and a wooden shoulder rest attached to the outside. The bazooka's "ammunition" was a small solid-fuel rocket, ignited by a simple electrical circuit connected to the trigger. The rocket could theoretically travel 400 to 500 yards, but was truly effective only at much shorter ranges: 120 yards or less. The rockets carried a special 3.5-pound "shaped charge" warhead capable of crippling a heavy tank or destroying a lighter armored vehicle, but they left a smoke trail that could betray the position of the launcher. Bazooka teams (one soldier aiming and firing, one preparing and loading rockets) thus required steady nerves. Like the crews of larger, vehicle-mounted rocket launchers, they had to master a rhythm of firing, moving, and firing again.

The bazooka was, by far, the most effective infantry antitank weapon of the war. It was used by U.S. armed forces in every theater, and exported to the Soviet Union for use by the Red Army. Partly in response, German tanks began to sport "skirts": vertical armor plates suspended along their sides to protect their vulnerable tracks and suspensions. The German army, meanwhile, studied captured bazookas and developed a very similar weapon nicknamed the *panzershreck* ("tank terror"). Slightly longer than the bazooka, it fired a rocket with a heavier warhead and a longer burning motor, which made it more effective at longer ranges. According to some reports, it could destroy stationary, lightly armored targets at 1,000 yards—ten times the effective range of the bazooka. The bazooka itself was steadily improved during the war. The M9 model, introduced in 1944, had an improved ignition system and a tube that could be broken down into two sections for ease of transport. It remained a short-range weapon, however; more powerful, longer ranged projectiles did not become available until after the war.

Direct-fire rockets launched from airplanes were a more efficient tool for destroying armored or reinforced targets. Bigger and heavier than their shoulder-launched counterparts, they could deliver a larger explosive charge. The standard British rocket (called the RP, for "rocket projectile") consisted of a 3-inch-diameter tube with four fins at the tail and a 65-pound, 5-inch-diameter warhead at the front. The standard American rocket from December 1943 on was the 5-inch FFAR ("forward-firing aircraft rocket"): essentially a rocket motor capped with a shell from a 5-inch antiaircraft gun. An upgraded version of the FFAR, using a larger motor and the same warhead, entered service in July 1944 under the designation HVAR ("high-velocity aircraft rocket"). Nearly twice as fast as its predecessor (485 versus 950 mph), it could penetrate 1.5 inches of steel armor or 4 feet of reinforced concrete. The exclamations of pilots startled by its power gave it a nickname: "Holy Moses." Even the 6-foot-long, 140-pound Holy Moses was dwarfed, however, by the ironically misnamed "Tiny Tim." Ten and a half feet long, nearly a foot in diameter, and weighing over 1,200 pounds, the Tiny Tim used a 500-pound armor-piercing bomb for a warhead and was designed for use against Japanese ships. Smaller rockets could be fired from beneath the wings of airplanes (the British RP from rails, the American FFAR and HVAR from stubby, streamlined pylons), but the Tiny Tim had to be slung beneath an airplane's belly and dropped free before its motor was ignited. One of the most powerful air-launched weapons of the war, it was deployed by the Navy in the Pacific but (apparently) never fired in anger.

Direct-fire rockets, because they could not be steered in flight, were of limited use against highly maneuverable targets—aircraft in flight, or motor vehicles moving on open ground. They were devastating, however, against vehicles that were dug into defensive positions, grouped into tight formations, or traveling in columns along roads. After the tide of the war in Europe began to turn in late 1942, German motorized units were frequently forced into such positions. The air forces of the three major Allied powers thus found themselves, by mid-1943, in position to use air-launched rockets to deadly effect. Two battles from that period will serve as illustrations here, but air-launched rockets were equally critical in dozens of others.

The battle of Kursk, on July 5–13, 1943, marked the turning point of the war in Russia, and the turning point at Kursk came on July 7. The Red Army and Air Force mounted a massive counterattack against advancing German forces, and the Illyushin Il-2 *Shturmovik* (Storm Bird)—a heavily armored airplane designed specifically for ground attack—played a decisive role. Armed with eight 82-milimeter rockets as well as cannon and bombs, the *Shturmoviks* took a heavy toll of German tanks and motor vehicles. Soviet reports credit one attack with destroying seventy tanks in twenty minutes, and a four-hour series of attacks with destroying 240 of the Seventeenth

Panzer Division's 300 tanks. The air assaults intensified over the next three days, and by July 10 the German forces were in disarray.

The battle of Mortain took place two months after D-Day, on August 7, 1944. The Allied attempt to break out of Normandy and press deeper into France was well underway, and Hitler had ordered his commanders to resist the breakout at all costs. German forces under the command of Gunther von Kluge counterattacked at the village of Mortain—the weakest place in the Allied lines—on that morning. Von Kluge had two infantry divisions and five armored divisions but *not* command of the skies above them. German fighters appeared over the battlefield, but were quickly shot down or driven off. By the time the German armored forces had been located, Allied ground-attack aircraft were able to fly mission after mission against them, unmolested. British pilots flying Hawker Typhoons devastated the German armor with rocket fire—eighty-three tanks destroyed, another twenty-nine probably destroyed, and twenty-four more damaged—blunting the attack and enabling Allied infantry to resist it. The result of the Typhoon attacks, Allied supreme commander Dwight Eisenhower later wrote, "was that the enemy attack was effectively brought to a halt, and a threat turned into a great victory" (Hallion 1989, 217).

Direct-fire rockets also proved effective against surfaced submarines. Even a relatively small rocket could, if fitted with an armor-piercing warhead, punch a hole in the sub's pressure hull and prevent it from submerging. Once trapped on the surface, the submarine could be captured or destroyed at leisure with bombs, guns, or more rockets. British antisubmarine rockets used the standard 3-inch body fitted with a 25-pound armor-piercing head instead of a 60-pound high-explosive one. The standard American rocket was the 3.5-inch FFAR—the beginning of a lineage that culminated in the Holy Moses. Pilots from Britain's Royal Navy executed the first successful rocket attacks on a submarine in May 1943, when a carrier-based Fairey Swordfish bomber damaged the *U-572*. After further aerial attacks and further damage, the U-boat's crew abandoned and scuttled their vessel. The Swordfish, ironically, was already obsolete when the war began. The last biplane to fight for any major combatant, it was given new potency by its high-tech rocket armament.

ROCKET PROPULSION

Choosing rockets as a propulsion system means trading range and endurance for acceleration and raw speed. Vehicles propelled solely by rockets are, therefore, impractical except when speed is absolutely critical—rising at a moment's notice, for example, to intercept incoming enemy bombers. The Allies had no real need for rocket fighters after 1942. The threat of German bombing had receded, and high-performance fighters had begun to roll off assembly lines in quantity. Germany and Japan, however, grew increasingly desperate as Allied forces advanced on them in 1944–1945. Both nations developed and deployed rocket-powered fighters as part of their increasingly desperate effort to avoid defeat.

The Messerschmitt Me-163B *Komet* was a rocket-powered interceptor designed to defend strategic targets against Allied bombers. Designed by German's most innovative aeronautical engineer, Dr. Alexander Lippisch, it was a small, single-seat airplane with a bomb-shaped aluminum fuselage, swept-back wooden wings, and no horizontal tail surfaces. Its liquid-fuel rocket engine was powered by two dangerously volatile chemicals: highly concentrated hydrogen peroxide stabilized with phosphate (a mixture called *T-stoff*), and a solution of hydrazine hydrate in methanol (*C-stoff*). When combined, even in tiny quantities, *T-stoff* and *C-stoff* ignited and burned with explosive force. Their power gave the *Komet* extraordinary performance: a top speed well over 500 mph (faster than any Allied fighter), an 80-degree angle of climb, and the ability to reach 40,000 feet in under five minutes. In the air, under rocket power, the *Komet* was literally unstoppable.

The *Komet*'s rocket motor was the key to its extraordinary performance, but also its greatest weakness. Both *T-stoff* and *C-stoff* were flammable, poisonous, and highly corrosive. Mixing them in the wrong proportions caused them to explode, and spilling them on anything organic caused it to dissolve or burst into flames. *Komet* pilots wore special protective coveralls, but even these were little protection in landing accidents where the plane overturned and the *T-stoff* tank behind the cockpit ruptured. The motor consumed fuel at a ferocious rate, giving the pilot a little over seven minutes of powered flight, and less than five at the altitude where the bombers flew. Its balky throttle mechanism discouraged pilots from changing speed while in flight, forcing them to attack their targets at speeds so high that accurate shooting was difficult. Once the *Komet*'s fuel was exhausted, it became a high-speed glider that was easy for Allied fighters to catch and shoot down. The most successful of the two squadrons placed in active service in 1944–1945 destroyed nine American bombers, but lost fourteen of its own fighters in the process. The other squadron scored no victories at all.

Axis leaders regarded the *Komet* as a first step on the road to safer, more effective rocket-powered interceptors. Me-163C and -D models, as well as a more sophisticated Me-263, were on German engineers' drawing boards when the war ended. Japan's own rocket fighters—the navy's J8M and the army's Ki-200—were near-duplicates of the *Komet*, built by copying plans and a sample rocket engine sent to Japan by submarine in 1943. Neither was operational by the time the war ended.

Japan did, however, deploy a unique rocket plane of its own design: the Yokosuka MXY7. Named the *Ohka* ("cherry blossom") by the Japanese and the *Baka* ("idiot") by U.S. naval intelligence officers, it was essentially a missile that used a human pilot as its guidance system. Built of wood and non-strategic metals, an *Ohka* had three solid-fuel rockets in its tail, over 2,600 pounds of high explosive in its nose, and a rudimentary cockpit in the space between them. It looked like a small, ugly airplane: a 20-foot cylinder with stubby, square-tipped wings and tail and a bulbous cockpit canopy. Dropped from beneath a specially modified twin-engine bomber as much as 20 miles from the target, it would glide toward its target on stubby wooden wings. The pilot, using airplane-style flight controls and a gun sight, would identify the largest ship in the immediate area, light the rockets, and dive into his target at speeds approaching 600 miles per hour. The *Ohka* was designed in late 1944, roughly the same time that the last of the Japanese fleet was being annihilated at the Battle of Leyte Gulf. It was intended as a last-ditch weapon for defending the Japanese coast against a seemingly inevitable Allied invasion, and reflected the Japanese military doctrine that one life was a small price to pay for the destruction of an enemy ship.

The rocket-powered *Ohka* was fast enough to outrun U.S. fighters and frustrate antiaircraft gunners, but only when the rockets were firing. When attached to the bombers that carried them, or when gliding toward their target, they were easy to shoot down. First used during the Battle of Okinawa in March 1945, they sank one destroyer (the USS *Mannert L. Abele*, on April 12, 1945) and inflicted varying amounts of damage on two other destroyers, a battleship, and several troop transports. It was an impressive showing for a makeshift weapon, but only a scratch on the massive U.S. fleet. The psychological effect of the *Ohka* was greater, but harder to measure. Because it was *designed* for suicide attacks, it reinforced the American perception of Japanese warriors as fanatics. Virtually impossible to shoot down during the final seconds of its flight, it was even more unnerving than a typical *kamikaze* aircraft. Anecdotal evidence suggests that, despite the limited damage it did, American sailors feared and despised it above all other Japanese weapons.

GUIDED MISSILES

Barrage rockets had been used in combat for centuries before World War II began. Direct-fire rockets had been used, albeit without much success, by the French Air Force in World War I. Guided missiles, however, were an entirely new development. They offered the advantages of rockets—speed, hitting power, light weight, low recoil—and eliminated rockets' most obvious flaw: inaccuracy. Even when aimed at a distant or moving target, guided missiles offered a reasonable chance of hitting it. The technology was still in its infancy when World War II ended, but it was already clear by 1945 that the new missiles had sown the seeds of a military revolution.

Britain and the United States experimented with guided missiles during the last years of the war, but on a relatively small scale. The missiles they did develop were built to meet specific battlefield needs. Both nations, for example, developed radio-guided missiles as a defense against Japanese *kamikaze* suicide aircraft: Britain's "Stooge" and America's "Little Joe" and "Lark." All three missiles used two sets of motors: solid-fuel "booster" rockets for takeoff and liquid-fuel "sustainer" rockets for flight. The missiles were, in effect, small remote-controlled airplanes "flown" toward the target by a sharp-eyed operator on the ground. The guidance system was far from precise, but the proximity fuse—an Allied innovation that exploded the warhead if it passed near the target—made "close" good enough. The U.S. Navy also developed two guided missiles designed to be carried by airplanes and launched against ground targets. Both the TDR-1 and the ASM-2 (nicknamed "Bat") were basically small gliders with large warheads mounted in their noses. The TDR-1 was "flown" by radio control by controllers riding aboard the airplane that carried it aloft. A forward-facing television camera relayed pictures of the approaching target during the final moments of the missile's flight, enabling the controllers to aim it more precisely. The Bat, in contrast, was fully independent once it was released from its carrier plane. Dropped at a height and heading that would cause it to glide to the target, it was kept on course by onboard gyroscopes. An onboard radar set, linked to the glider's control surfaces, bounced radio waves off the target as the missile approached, and automatically supplied final course corrections.

None of the Allies' antiaircraft missiles saw frontline service during the war. TDR-1s and Bats saw limited use and, within those limits, considerable success. Deployed by the U.S. Navy in the Pacific, they sank Japanese ships and destroyed bridges and antiaircraft sites. Germany achieved even greater success with its own guided antiship missiles: the Hs-293 and the larger but less accurate RD-1400, better known as the "Fritz-X." Both the Hs-293 and the Fritz-X were radio-controlled gliders that were steered toward their targets by an operator aboard the plane that dropped them. Each carried a small rocket engine to accelerate it during the final seconds of its flight, maximizing both the chances of a hit and the damage a hit would do.

Introduced in August 1943, the Fritz-X scored a series of spectacular successes in the Mediterranean: sinking the Italian battleship *Roma* and the British cruiser *Spartan*, and damaging the Italian battleship *Italia*, the British battleship *Warspite*, and the American cruisers *Philadelphia* and *Savannah*. The Hs-293 also enjoyed a string of early successes, sinking a number of transports and small warships. The effectiveness of both weapons gradually declined, however, as the Allies learned to use electronic jamming equipment to disrupt their guidance systems.

The most significant German missiles of the war, however, were designed not for use against ships but for use against cities. Hitler dubbed the V-1 and V-2 *Verstellungswaffe* ("vengeance weapons"), and saw them as a means of terrorizing Allied civilians and so destroying their will to fight. The mission that Hitler envisioned for the V-1 and V-2 was essentially the same one that German bombers had carried out against Britain during the "Blitz" of 1940–1941. There was, however, one critical difference. Bombers could be shot down or turned aside by fighters and antiaircraft guns, but the V-1 was (initially) difficult to stop and the V-2 could not be stopped at all.

The V-1 (officially the Fiesler F-103) was a small, unpiloted airplane powered by a jet engine and guided by a system of gyroscopes linked to its rudder and elevators. The V-1 was designed for mass production. The wings and fuselage were made of sheet metal, the engine was a simple "pulse jet" (little more than a carefully shaped tube with a fuel injector and an igniter), and the ingenious guidance system was built simply and from off-the-shelf hardware. Thirty thousand V-1s were built in all, and between June 1944 and March 1945 10,000 were fired at England from launch sites on the coasts of France and Holland. Seven thousand fell on English soil, a little over half in London and its suburbs. The V-1s had their greatest impact in the summer of 1944. They flew too fast for antiaircraft gunners to shoot down or for most fighters to catch, and announced their coming with a loud, distinctive buzzing sound that gave them the nickname "Buzz Bomb." The noise meant that (as Hitler had intended) the missiles sowed fear and anxiety even in areas where they did not fall. Londoners who lived through the summer of 1944 recalled, after the war, the way that they would cock their ears when they heard the rising buzz that signaled a V-1's approach. Life stood momentarily still until the bomb passed over and the buzz began to fade again, or until the sound of a distant explosion signaled that it had hit somewhere else. "Most of the people I know," wrote Harry Butcher, a senior American officer stationed near London, "are semi-dazed from loss of sleep and have the jitters, which they show when doors bang or the sounds of motors from motorcycles to aircraft are heard" (Irving 1982, 171).

The success of the V–1 diminished rapidly, however, after mid–August 1944. Their launch sites were pounded by Allied bombers and, in time, overrun by allied troops. High–performance fighters were rushed to southern England, and their pilots gradually developed techniques of destroying the missiles. The most effective defenses against the V–1, however, came in the form of two critical upgrades to antiaircraft guns. The first, a radar aiming system, made it easier for gunners to track the fast-moving V-1s. The second, the same proximity fuses later used on Allied guided missiles, increased the chances of a lethal hit by causing shells to explode when they passed close to the target. The V-1 bombardment went on for seven more months, but the vast majority of the missiles were shot down before they reached London.

The V-2 was a far more sophisticated weapon than the V-1 and, therefore, a far greater problem for the Allies. It was the world's first operational ballistic missile, designed to be launched vertically and soar to the top of a high arc before falling toward its target. Developed by a team led by Walter Dornberger (Karl Becker's assistant) and Wernher von Braun, the V-2 was a development of the A-2 and A-3 rockets the team had developed in the late 1930s. The V-2 was powered by a single-chamber rocket motor. A pump near the tail, turned by gas generated the decomposition of hydrogen peroxide, fed steady streams of alcohol (the fuel) and liquid oxygen (the oxidizer) into the chamber. Between the tanks and the engine, the fuel circulated around the outside of the chamber in tubes—an ingenious design that warmed the fuel (to make it easier to ignite) and cooled the chamber (to keep it from melting). The alcohol and liquid oxygen entered the chamber through small holes in an "injector" (initially shaped like a flat plate, later like an inverted cup), which turned them from streams of liquid to easy-to-ignite clouds of tiny droplets. The V-2 was kept on course by a guidance system consisting of gyroscopes and a primitive analog (mechanical) computer. Movable graphite vanes, moving according to the computer's commands, steered the missile by deflecting the exhaust stream to one side or another. Range was controlled by putting more or less fuel in the tanks before launch: when the fuel was exhausted, the V-2 would stop climbing and begin its supersonic plunge to Earth.

The V-2 was a far-from-precise weapon. It could reliably hit city-sized targets, but not much more. There was no question of singling out a particular factory or military base. It was a blunt instrument, designed to kill people and destroy property at random. Its psychological effect was the opposite of the V-1's. Rather than announcing its arrival with a noisy buzz, it fell onto its target silently, its fuel expended and its engine cold. Residents of areas hit by the V-2 typically knew they were under attack only when the

Figure 4.2: Homes reduced to rubble by a V-weapon explosion near
Camberwell Road, London, in 1944. Office of Strategic Services photograph.
Courtesy of the National Archives at College Park, Maryland, War and
Conflict image number 1324.

explosions began. The speed of the V-2 made it impossible to shoot down. The only way to stop them was to destroy their launch sites, and after several elaborately prepared concrete bunkers were bombed into ruin, V-2 units (part of the artillery arm of the German army) adopted a radically different strategy. Using specially prepared trucks and trailers, they transported the missiles by road and fired them from portable metal launch stands. A convoy carrying everything necessary to launch a V-2 might involve as many as thirty trucks, but it *was* mobile. Air strikes were of little use against a "launch site" that could be picked up and driven onto nearly any road in occupied Europe.

Germany produced just over 6,000 V-2s between 1942 and 1945, most of them on assembly lines in underground factories manned by slave laborers. Roughly 3,400 of them were fired at Allied targets and nearly 2,900 of those struck home, with 1,500 falling on London alone. More than 2,500 more V-2s were captured by Allied troops when they overran the launch sites and factories in the spring of 1945. Most were destroyed, but others were exported (along with members of the design team) to the United States and Soviet Union.

The damage done by the V-1s and V-2s was substantial: 33,700 buildings destroyed and 204,000 damaged; 12,685 people killed and 26,433 injured (see Figure 4.2). It paled, however, beside the damage that Allied bombers were capable of inflicting in 1944–1945. The bombing of Dresden (February 13–15, 1945) and Tokyo (March 9–10, 1945) each wrought more destruction in a day than the entire V-weapons program did in a year. The V-1 (made of sheet metal and fueled by gasoline) was relatively cheap, but the V-2's use of scarce aluminum, alcohol, and liquid oxygen made it expensive. Each V-2 cost nearly as much as a fighter plane, and developing the program cost Germany as much (in relation to the size of its economy) as developing the atomic bomb cost the United States. Viewed in terms of military efficiency, the V-2 was a failure. Each of the 1,500 missiles that landed on London killed an average of only 1.76 people.

The V-1 and V-2 were critical, however, in proving that cruise missiles (like the V-1) and ballistic missiles (like the V-2) were technologically viable weapons. The Cold War, already taking shape in March 1945 as the last V-weapons lifted off, would be "fought" with their direct descendents.

5

Rockets for Research, 1945–1960

The advantages of rocket power were clear by the end of World War II. First, rocket motors could supply massive amounts of thrust almost instantly. Second, they could accelerate payloads to speeds that no piston or jet engine could match. Finally, they were relatively simple and lightweight. The principal disadvantage of rocket power was equally clear by 1945. Rocket motors could not yet be throttled—they ran at full power or not at all—and so exhausted their fuel in a matter of minutes. Rocket power was, therefore, suitable only for vehicles designed for short, high-speed runs: short-range interceptors like the *Komet* and guided missiles like the V-2 and *Ohka*. A conventional fighter powered by rockets would be hopelessly impractical. A bomber or commercial transport could use them efficiently only by climbing to the edge of space and gliding through the upper atmosphere to reach destinations on the far side of the world.

Even before the war ended, however, a new application for rocket power began to emerge: research into high-speed flight and atmospheric science. Wartime demands led, between 1939 and 1945, to rapid improvements in aircraft performance. The ability to fly higher and faster than ever before demanded a better understanding of how pilots, airplanes, and the air itself behaved at high speeds and high altitudes. It also spurred interest in new types of engines that would expand the performance "envelope" even further. The United States led the world in high-speed, high-altitude research in the two decades after World War II. Rockets were central to that research program, propelling instruments into the upper atmosphere and aircraft (both piloted and unpiloted) to new speed and altitude records.

Rocket motors offered significant advantages as a power source for research aircraft. No other power source could reliably carry an aircraft and its instruments to such high speeds and high altitudes. Rocket motors also offered versatility. They could be removed from the aircraft, and replaced with new motors offering better performance, more easily than jets could. Rocket motors could also be configured as self-contained "strap-on" packages and attached to jet-powered aircraft in order to improve their performance for testing purposes. The nature of experimental flight research also minimized rocket planes' drawbacks. The careful planning that preceded each flight enabled pilots and engineers to take rocket motors' brief endurance into account. The rhythm of research programs—plan, fly, analyze, repeat—gave ample time for the rocket planes to be refurbished and reattached to the converted bombers that carried them aloft. Impractical for production military or civilian aircraft, rocket power found a long-term home in flight research programs.

ROCKET PLANES AND THE "SOUND BARRIER"

The Bell X-1 was the first rocket-powered research plane in history. It remains the most famous, and with good reason. It demonstrated that flying faster than the speed of sound was possible and, in a properly designed aircraft, safe. Neither seemed a foregone conclusion in 1944, when the U.S. Army Air Force inaugurated what would become the X-1 program in 1944. Piston-engine aircraft capable of approaching the speed of sound in a steep dive were already in service by 1944, and their pilots had reported severe buffeting and loss of control at such high speeds. Some aeronautical engineers speculated that it might not be possible for an aircraft to reach the speed of sound without losing control or breaking up, and the idea of a "sound barrier" entered popular culture. The Army Air Force, the Navy, and the National Advisory Committee on Aeronautics (NACA, the forerunner of NASA) all pursued research on transonic flight. All three concluded that buffeting and control problems would diminish at speeds above that of sound, and that a properly designed aircraft could survive it. Both armed services began programs to build and fly such an aircraft, and both (against NACA recommendations) eventually chose rocket motors for propulsion.

Designed and built at the Buffalo, New York, headquarters of Bell Aircraft in 1945, the X-1 was unlike any American aircraft before it. The fuselage was shaped like a .50-caliber machine gun bullet: an object known to be stable at supersonic speeds. The wings were thin and flat, and the entire horizontal tail functioned as a movable control service (like the elevators of a conventional aircraft). The X-1 was powered by a single, four-chamber, alcohol-oxygen rocket motor designed by James Wyld and built by Reaction Motors, Inc. The rocket could not be throttled in flight, but each of the four combustion chambers could be ignited independently, giving the pilot some control over its thrust. The original design called for a pump to feed fuel into the chambers, but manufacturing problems forced the designers to use pressurized nitrogen gas instead. The replacement system was simpler than the pump—nitrogen, pumped into the alcohol and oxygen tanks under pressure, would force the fuel out—but it was heavier and (because the nitrogen had to be stored in a tank of its own) bulkier. The change meant that the X-1 could carry less fuel, and fly for shorter times, than designers had originally hoped. It also eliminated any possibility of the X-1 taking off and climbing to altitude under its own power. Even with these limitations, however, the first "X plane" proved itself more than equal to the job it had been designed to do.

Bell delivered two X-1s to Muroc (now Edwards) Air Force Base in 1946. Both were extensively tested by company pilots, both in glides and in powered flights up to 80 percent of the speed of sound (Mach 0.8). NACA and the Air Force agreed, at a June 1947 meeting, to use the two aircraft to carry out simultaneous, complementary research programs. The Air Force, using the first X-1, would focus on achieving supersonic speeds. NACA, using the second, would investigate stability and control at supersonic speeds. Both programs gathered valuable data, but it was the Air Force's speed runs that captivated the public. In a series of twelve flights between early August and early October 1947, Captain Charles ("Chuck") Yeager took the X-1 steadily closer to the speed of sound. The twelfth flight of the series, on October 10, reached Mach 0.997. Four days later, on October 14, Yeager became the first pilot in history to travel faster than sound in level flight. Released from a B-29 bomber at 20,000 feet, he fired two of the rocket motor's four chambers and climbed to 40,000 feet to begin the test. Firing the third chamber, Yeager felt the X-1 accelerate rapidly and saw the "Mach meter" on the instrument panel spin to, and then past, Mach 1. At Mach 1.02, he wrote in his flight report, "the meter momentarily stopped and then jumped to 1.06 and this hesitation was assumed to be caused by the effect of shock waves" (Miller 1988, 19). Yeager stopped the engines, allowed the X-1 to decelerate to subsonic speed—noting a single "bump" at Mach 0.98—and glided down to an uneventful landing. The "sound barrier" had proved to be no barrier at all, and the realm of supersonic flight stood wide open.

NACA and the Air Force pushed their X-1s further and further into that realm over the next few years. The two aircraft were retired (number 1 in May 1950, number 2 in October 1951) with a total of 157 flights to their credit. The flights revealed that the top speed of the "first-generation" X-1s was about Mach 1.45, and that their maximum ceiling was about 70,000 feet. The three "second-generation" X-1s that entered Air Force service beginning in 1953 significantly expanded these capabilities, routinely reaching speeds above Mach 1.5 and, in a series of flights in mid-1954, altitudes above 87,000 feet. The Bell X-2, designed with swept wings and a more powerful rocket motor, expanded the performance envelope even further in 1955 and 1956. Colonel Robert Everest pushed it past Mach 2.5 in July 1956, and Captain Iven Kinchloe reached an altitude of nearly 126,000 feet in September of the same year. Three weeks after Kinchloe's flight, Captain Milburn Apt became the first pilot to exceed Mach 3, but was killed when he lost control of the X-2 during its gliding descent and failed to eject in time.

The Navy's research on supersonic flight ran concurrently with the Air Force's between 1947 and 1956. The Navy's research planes, built by Douglas Aircraft in California and designated D-558, used a more conservative design than their Air Force counterparts. The first three—the D-558-1 Skystreak series—had straight wings, a single jet engine, and maximum speeds below Mach 1 in level flight. The second three—the D-558-2 Skyrocket series—used swept wings and two engines: a jet for takeoff and low-speed flight and a Reaction Motors rocket for achieving and maintaining supersonic speeds. The second Skyrocket was heavily modified in 1950: the jet engine was removed and replaced by additional fuel tanks for the rocket, enabling it to reach higher speeds and sustain them for longer periods. From 1950 to 1953, when the first-generation X-1s had been retired and the second-generation X-1s had not yet entered service, it was the nation's premier supersonic research aircraft. Douglas, Navy, and NACA pilots set a series of speed and altitude records in it, and in 1953 Scott Crossfield became the first pilot in history to exceed Mach 2.

The hundreds of X-1, X-2, and D-558 test flights made between 1947 and 1957 defined the problems that designers of supersonic aircraft would have to solve. The planes themselves introduced technological innovations that would become standard in production supersonic aircraft. The X-1, for example, showed that thin, sharp wings and movable horizontal tail surfaces could significantly reduce buffeting at speeds around Mach 1. The D-558-2 pioneered the use of titanium as a heat-resistant structural material, and the X-2 did the same with specialized forms of stainless steel. Both aircraft incorporated ejection systems that separated the entire pilot's cabin from the rest of the aircraft, using it as an "escape pod" to protect the pilot until he could bail out manually. When the first supersonic warplanes—the F-100 Super Saber fighter and B-58 Hustler bomber—began to enter military service in the late 1950s, they bore an unmistakable family resemblance to the rocket-powered research planes that had preceded them.

AMERICAN HIGH-ALTITUDE ROCKETS

The U.S. military's interest in rocket propulsion did not stop with supersonic flights over the California desert. Both the Army and the Navy also developed and tested "sounding rockets" designed to carry payloads of instruments. "Sounding" is an old sailor's term: the process of measuring the depth of water and, originally, the composition of the sea bottom directly beneath the ship. The sounding rockets of the late 1940s and 1950s were used to carry out similar explorations of the upper atmosphere. They carried instruments that measured temperatures, winds, and radiation levels, and eventually used cameras to return the first high-altitude photographs of storms and other meteorological phenomena.

The first sounding rocket, the WAC Corporal, was designed by Frank Malina and built by the Jet Propulsion Laboratory (JPL) in less than ten months. The first design studies began in December 1944, and the first test flight took place at White Sands, New Mexico, in September 1945. The WAC Corporal was intended as a technological stepping stone between two other missiles: the small, solid-fueled Private and the larger, liquid-fueled Corporal. Building and flying it would, JPL director Theodore von Karman believed, give the lab much-needed experience with the still-experimental technology of liquid-fuel engines before they tackled the Corporal itself. Sixteen feet tall and weighing 700 pounds, the WAC Corporal used a main engine designed by Aerojet that burned aniline fuel with nitric acid as an oxidizer to create 1,500 pounds of thrust. It carried no guidance system and had only its tail fins for stability (WAC officially stood for "without attitude control")—a design that made rapid acceleration critical. A strap-on solid-fuel booster (actually a "Tiny Tim" rocket obtained from the Navy) provided 50,000 pounds of thrust in the first few seconds of the flight, ensuring that the air flowing over the WAC Corporal's fins would be moving fast enough for them to do their job.

WAC Corporal's performance, though modest, far exceeded its design-ers' expectations. Expected to reach 100,000 feet on its first test flight, it topped out at over 230,000 feet—more than 40 miles above the Earth. It would likely have become the Army's principal high-altitude research vehi-cle, had the more capable V-2 not become available in early 1946. Even the arrival of the V-2, however, did not end the WAC Corporal's career. Eight of the little missiles were adapted in 1949–1950 for use as second stages on V-2s—the first operational test of the multistage concept pioneered by Tsi-olkovsky and Goddard decades earlier (see Figure 5.1). The fifth launch in what became known as Project Bumper reached a record altitude of just under 250 miles on February 24, 1949. Its WAC Corporal upper stage thus became the first rocket to (briefly) enter outer space. The last two V-2/WAC Corporal combinations, used to evaluate near-horizontal flight paths, were launched from the Army's missile test range at Cape Canaveral, Florida, in 1950—the first departures from what became America's first spaceport.

The payload of the WAC Corporal was limited to about 25 pounds. The V-2 could carry more, but tended to tumble once its engine stopped firing. Serious high-altitude research demanded larger payloads and a more stable platform, and the coalitions of scientists and military officers set out to create a rocket that could supply them.

The first solution was the Aerobee, designed and built—as the WAC Corporal had been—by Aerojet General and Douglas Aircraft. The pro-gram was funded by the Navy through its Bureau of Ordnance and Office of Naval Research, and overseen by the Applied Physics Laboratory at Johns Hopkins University—an organization that became for the Navy what JPL was for the Army. The Aerobee was essentially a larger and more capa-ble WAC Corporal. Nineteen feet long and weighing 1,600 pounds, its liquid-fuel motor generated 2,400 pounds of thrust—more than 50 percent better than its predecessor. Aerobee's performance was also a major step be-yond that of the WAC Corporal: it could lift up to 150 pounds of instru-ments to altitudes approaching 60 miles. The first fully operational Aerobee flew at White Sands in late 1947, and forty more were launched over the next eight years. The Aerobee-Hi, an improved model with more than dou-ble the performance of the original, entered service in the mid-1950s.

Viking, another Navy sounding rocket, was conceived in 1946 as a larger and more powerful complement for Aerobee—initially capable of carrying 500 pounds to an altitude of 100 miles. The development of Viking, like that of Aerobee, was overseen by and funded through the Naval Research Laboratory. Like many late-1940s rockets, however, Viking was really the product of wide-ranging collaboration. The project was overseen

Figure 5.1: A converted V-2 with a WAC Corporal upper stage, part of Project Bumper, lifts off from the Army Missile Range at Cape Canaveral, Florida, on July 24, 1950—the first launch from what would become Kennedy Space Center. Courtesy of NASA Kennedy Space Center, image number 66P-0631.

by Milt Rosen, an electrical engineer who took a seven-month crash course in rocket engineering at JPL. Martin, a leading aircraft manufacturer, built the rocket body, and a team of Reaction Motors engineers led by John Shesta designed the motor. Albert Hall of the Massachusetts Institute of Technology provided the mathematical basis of a new type of guidance system that allowed more precise and accurate in-flight control. Viking was influenced by the V–2, but took important steps beyond it. It was built of lightweight aluminum instead of heavy steel, and saved weight by using the outer skin of the rocket as the outer walls of the fuel tanks. Whereas the V–2's steering mechanism had used carbon vanes to deflect the exhaust stream, Viking had a radically new "gimbaled engine" that could swivel back and forth. The first three launches in the Viking program ended in failure due to problems with the engine, but seven of the next nine were complete successes. On a flight in May 1954, an upgraded version of Viking lifted 825 pounds of instruments to an altitude of 158 miles (see Figure 5.2).

The WAC Corporal, Aerobee, and Viking flights significantly advanced American scientists' understanding of the upper atmosphere and the edges of outer space. They also provided a unique new perspective on the Earth. Equipped with cameras and reinforced containers to return the film to Earth, they provided the first images of the Earth as seen from the edge of space. Their ability to see hundreds or thousands of miles "over the horizon," and to take in entire weather systems at a glance, hinted at what Earth–orbiting satellites might be able to accomplish. The sounding rockets' contributions to engineering were even greater. They gave firms like Reaction Motors and Aerojet, organizations like JPL and APL, and individuals like Milt Rosen and John Shesta valuable experience in building large liquid–fuel rockets. They pioneered design features that would become standard in later years: multiple stages, gimbaled engines, and integral fuel tanks. The sounding rockets were, technologically and operationally, a critical step on the road from the V–2 to the future. They set the stage for the ICBMs and satellite launchers of the late 1950s, just as the X–planes did for supersonic jets.

SOVIET HIGH-ALTITUDE ROCKETS

The Soviet Union developed its own stable of high–altitude rockets in the late 1940s and 1950s. Known collectively as "geophysical rockets," they were mostly adaptations of military guided missiles rather than purpose-built designs like the Aerobee. The R–1A, for example, was a slightly modified

Figure 5.2: The U.S. Navy's eleventh Viking sounding rocket lifts off from White Sands Proving Ground, New Mexico, on May 24, 1954. It set a new world altitude record of 158 miles. U.S. Navy photograph. Courtesy of the Library of Congress, image number LC-USZ62-108192.

version of the R-1 ballistic missile, which was itself a slightly modified, Soviet-built version of the V-2. The principal purpose of the R-1A was to test a new system for separating the payload section from the missile while in flight, a design that would enhance the striking power of military missiles. The first four of the six R-1As launched in May 1949 carried only dummy instrument packages. Only the fifth and sixth flights, flown after the separable payload module had been deemed a success, carried actual scientific instruments. These flights reached altitudes of roughly 60 miles— equivalent to the original Aerobee—taking air samples and measuring atmospheric temperature and pressure. Several of the larger, more capable ballistic missiles that followed the R-1—notably the R-2 and R-5—had variants designated for geophysical research. The R-2, used throughout the 1950s, could routinely boost payloads to altitudes of more than 125 miles— slightly better than the performance of the U.S.-built Viking. The R-5, the first Soviet missile with a range exceeding 1,000 miles, could do even more. Over a series of four flights in 1958, they carried scientific payloads to altitudes of 282 miles, a record for single-stage missiles.

Atmospheric observation continued to be a priority for Soviet space scientists in the 1950s. A report drawn up in 1951 by the Commission for the Investigation of the Upper Atmosphere laid out an eight-point program that could be accomplished with existing R-1 missiles. It included investigations of wind velocities, radiation levels, and the chemical composition of the air at high altitudes. One innovative experiment from the program used a rocket that set off smoke flares at predetermined altitudes, allowing scientists to track upper-atmosphere wind patterns from the ground. Significantly, the report also called for research into animal behavior at high altitudes. Plans to launch humans into space had been part of the Soviet rocket program since at least 1948, and putting animals aboard high-altitude research rockets was a first step toward that goal.

The first "crews" to fly on Soviet high-altitude rockets were dogs: nine animals chosen for their small size, trainability, and light-colored coats (to facilitate photography inside the poorly lit payload compartment). A newly formed community of Soviet space medicine experts rounded up nine dogs, and between July and September 1951 a series of six R-1 flights lofted them two at a time into the upper atmosphere. Four of the nine dogs died in landing accidents, but overall the program was a success. The animals suffered no ill effects from traveling at 2,500 miles per hour to altitudes of 60 miles or more, and good-naturedly accepted the several minutes of weightlessness that each flight involved. A second series of R-1 flights in 1954 and 1955 put twelve dogs (including four "veterans" of the 1951 flights) through more elaborate medical tests. It also equipped them with specially designed spacesuits, and used an automated system that ejected one dog from each "crew" during the descent phase of the flight. The ejected animal, clad in its spacesuit, descended to Earth on its own parachute, while its partner rode to Earth in the payload container. The final series of vertical-trajectory dog flights, carried out in 1957 and 1958 using R-2 and R-5 missiles as launch vehicles, used simpler flight plans. They doubled and redoubled the altitudes reached by the earlier flights, however, taking their canine passengers beyond the edge of the atmosphere.

The Soviets' high-altitude rocket program did not, as the American program did, serve as a testing ground for new concepts in rocket design. Instead, it used derivatives of well-tested military missiles to carry its payloads. It *did* break new technological ground, however, in the design of life-support systems, spacesuits, medical monitoring equipment, and crew-recovery techniques. That knowledge became, in 1959–1961, the basis of some of the Soviet Union's most spectacular achievements in space.

ROCKET PLANES AND SPACE

The X-15, conceived by the National Advisory Committee on Aeronautics (NACA) in 1952 and built by North American Aviation in 1955–1958, was the ultimate rocket-powered research airplane. Like the X-1, X-2, and D-558, it was designed to investigate high-speed, high-altitude flight. Sponsored jointly by the Navy, the Air Force, and NACA, it was designed both to gather scientific data and to test design features that could be incorporated into future aircraft and spacecraft. Its operating environment would be far beyond that of any previous research aircraft: speeds of Mach 4 to Mach 6, and altitudes in excess of 200,000 feet. The X-15 was, as a result, the world's first aerospace plane, capable of operating both in the Earth's atmosphere and on the fringes of outer space.

The X-15's XLR99 rocket motor—the key to its unprecedented performance—was designed and built by Reaction Motors, by now an established supplier of liquid-fuel rockets. The basic elements of the motor were familiar from earlier designs: alcohol for fuel, liquid oxygen for oxidizer, and pumps to supply them to the combustion chamber. The XLR99, however, introduced two critical innovations. It was the first large liquid-fuel rocket motor that could be "throttled" and the first that could be stopped and then restarted while in flight. Throttling, which allowed the pilot to vary the motor's thrust from 30 to 100 percent of full power, was accomplished by the simple-but-effective method of varying the speed of the fuel-supply pumps. The pumps themselves were connected to a single shaft with a turbine at one end. Oxygen gas, created by passing liquid hydrogen peroxide over a catalyst, emerged from a nozzle and sprayed across the turbine blades, spinning the shaft and working the pumps. Increasing the flow of hydrogen peroxide increased the flow of gas, the speed of the pumps, and so the rocket motor's thrust.

Flown by pilots from the Air Force, the Navy, and the National Aeronautics and Space Administration (which supported NACA in October 1958), the three X-15s built by North American made a total of 199 flights between June 1959 and October 1968. The later flights in the program routinely exceeded Mach 5, and two—flown by Captain William Knight in November 1966 and May 1967—exceeded Mach 6. More than a dozen X-15 flights exceeded 250,000 feet (the altitude at which Earth's atmosphere officially ends and space begins), earning their pilots the right to wear astronaut wings on their uniforms. The X-15 pioneered the use of high-strength, heat-resistant structural materials and specially formulated heat-resistant "paint" (actually a mixture of resin, catalyst, and tiny glass beads). It was the first aircraft designed with two full control systems: aircraft-style control surfaces for operations in the dense lower atmosphere, and gas jets for the thin upper atmosphere and airless space. It was also the first that could pass out of, and back into, the Earth's envelope of air. The men who flew the X-15 were the first airplane pilots to wear full pressure suits, and the first to experience weightlessness for more than a few seconds. They explored the boundary between air travel and space travel, much as the pilots of the X-1 and X-2 explored the boundary between subsonic and supersonic flight (see Figure 5.3).

The X-15 program began in the mid-1950s, when winged rockets seemed to be a logical—even obvious—solution to the problem of putting humans in space. Reaching space seemed, at the time, to be a matter of flying ever higher and ever faster until you left Earth's atmosphere and gravity behind. The X-15 seemed, when it began flying in 1959, to be a major step toward that goal. Military test pilots were drawn to the program, in part because it seemed to represent the future of aerospace flight: a true "spaceship" rather than the "space capsule" that would be used in NASA's just-announced Project Mercury. The commander of the Air Force's flight test center at Edwards Air Force Base actively discouraged his pilots from applying for Project Mercury. The Mercury "astronauts" would, the gossip at Edwards said, be nothing more than guinea pigs for biomedical experiments—"Spam in a can."

The commandant of Edwards was wrong, however. It was "capsules" that carried American astronauts (including former X-15 pilot Neil Armstrong) to orbit during the Mercury and Gemini programs and to the

Figure 5.3: NASA test pilot Neil Armstrong, who went on to command the first manned lunar landing in July 1969, poses next to the X-15 research plane in 1959. A hybrid of airplane and rocket technology, the X-15 was equally capable of operating in Earth's atmosphere and in outer space. Data gathered by Armstrong and other X-15 pilots were critical to the design of the space shuttle. Courtesy of NASA Dryden Flight Research Center, image number E60-6286.

moon during Project Apollo. The name "capsule" stuck, but by the time Project Gemini began in 1965 they had become true spacecraft, just as capable of being "flown" in space as the X-15. A second-generation X-15, capable of reaching orbit atop a cluster of modified Titan missiles, was designed but never built. The X-20 Dyna-Soar, an Air Force spaceplane that would ride a single Titan to orbit and glide back to Earth, was cancelled in 1961 before the prototype was completed. Unknown to American engineers, a single-seat Soviet rocket plane dubbed the PKA (a Russian acronym for Space Gliding Apparatus) had suffered a similar fate less than two years earlier. Large-scale models of it had been completed in 1959, but political support had failed to materialize and wind-tunnel tests had revealed unforeseen problems with its thermal protection system. Like the Dyna-Soar and the second-generation X-15, it was quietly shelved—a victim of aerospace technology's rapid movement in a different direction.

The original X-15 program continued, however, operating in the shadow of Mercury, Gemini, and Apollo and amassing reams of valuable data that would help to shape their successor: the space shuttle. The 200th and final flight of the program, postponed by bad weather and eventually cancelled, would have taken place in December 1968—just days before *Apollo 8* became the first spacecraft to fly beyond Earth's orbit.

6

Ballistic Missiles and the Cold War, 1945–1990

◆

The Cold War, like the two World Wars that preceded it, left millions dead and millions more permanently scarred in body and mind. Competition between the United States and the Soviet Union spawned wars, revolutions, guerilla campaigns, assassinations, and political executions throughout Asia, Africa, and Latin America. Countries barely touched by the World Wars—Chile, Angola, Iran, Afghanistan, Cambodia—were devastated by the "low-intensity" conflicts that the superpowers waged between 1945 and 1990. The Cold War was only "cold" in comparison to "World War III": a full-scale Soviet-American war that many feared would leave both nations (and much of the rest of the world) in ruins. Many politicians argued at the time, and some historians have argued since, that the Cold War was the price the world paid for avoiding World War III.

The existence of guided missiles capable of carrying nuclear warheads was part of what made the prospect of World War III so terrifying. Nuclear missiles could be based almost anywhere, launched on relatively short notice, and devastate a city in the blink of an eye. Once in flight they were beyond human control: those who had launched them could not call them back, and those they were launched at could not stop them. The existence of nuclear missiles created a new kind of nightmare: that a political or military misjudgment would spiral out of control, leading to a war that would devastate half the world. Soviet and American leaders alike thus considered their actions—and their enemy's responses—with the greatest of care, lest they commit such a misjudgment. Both superpowers also used their nuclear missiles for political leverage, each quietly but explicitly threatening prompt and total destruction of the other if its homeland or its vital interests abroad were threatened. Finally, both superpowers employed their missiles in more benign forms of competition: launching satellites, space probes, and human crews into orbit, for example. The value of ballistic missiles—as weapons, political tools, and objects of national pride—became evident to other nations as well, and as the Cold War ground on many set out to buy, borrow, or build missiles of their own.

THE FIRST BALLISTIC MISSILES

Germany's V-2, the first ballistic missile used in combat, could carry a 1-ton warhead a little over 200 miles. It could hit a city-sized target, but its CEP[1]—a rough measure of accuracy—was measured in miles rather than yards. The first Soviet and American ballistic missiles of the Cold War used the V-2 as a starting point, but steadily improved on its performance. The Soviet R-1 of 1948 (which western analysts designated SS-1 and code-named "Scud") was a straightforward copy of the V-2. The R-2, which first flew in 1949, had the same basic design but double the range, thanks to improved engine design and reformulated fuel. The R-5 (which the West knew as the SS-3 "Shyster") was the last direct descendant of the V-2 to be built in the Soviet Union. Introduced in 1956, it had four times the range of its German "ancestor" (700 miles) and could carry 50 percent more payload: 3,000 pounds instead of 2,000. Redstone, designed for the U.S. Army by Wernher von Braun and his team of German expatriates, was also closely modeled on the V-2. First flown in 1953, it was intended (like the V-2) to be carried by army truck convoys and fired from portable launch pads set up on open ground. Its range (200 miles) was comparable to that of the V-2, but it was more powerful and more accurate, capable of carrying a 7,000-pound warhead and delivering it with a CEP of roughly 1,000 feet. The test flights of the Redstone coincided with the first tests of a new generation of compact, lightweight nuclear bombs. Redstone thus became the first missile in either superpower's arsenal that was capable of delivering a nuclear warhead. The R-5M, an upgraded R-5 that first flew in 1955, gave

---

1. CEP stands for "circular error probable." It is the radius of a circle, centered on the target, within which 50 percent of the missiles fired at a target will land. The lower a missile's CEP, the more accurate the missile.

the Soviet Union the same capacity. The new missiles were, in a sense, the ultimate form of artillery: self-guided "shells" that could travel hundreds of miles and explode with the force of a million tons of high explosive.[2]

The R-5M and Redstone divided the first generation of ballistic missiles from the second. Their basic designs were firmly rooted in the technology of the mid-1940s, but they introduced features that became standard on second-generation missiles. Three innovations—gimbaled engines, inertial guidance, and separable warheads—were especially critical. Gimbaled engines, introduced on the sounding rockets of the late 1940s and also used on the R-5, offered more precise control with less reduction of thrust than the movable vanes used on the V-2 and Redstone. Inertial guidance used onboard instruments to detect changes in the motion of the missile and feed information about the direction and magnitude of the changes to the autopilot, which used the information to steer the missile back on course. It enhanced accuracy while eliminating the use of signals from the ground that an enemy might jam. Separable warheads, which detached from the main body of the missile at the peak of its trajectory and fell toward their target independently, simplified missile design by making the main body of the missile disposable. Only a robust machine could withstand a high-speed fall through the atmosphere. Separable warheads meant that only the "reentry vehicle" that protected the warhead itself had to meet those high standards of durability. Separable warheads also increased accuracy, since a specially designed reentry vehicle was easier to stabilize during its long fall to Earth than a whole missile would be.

The second generation of ballistic missiles used gimbaled engines, inertial guidance, and separable warheads as a matter of course. They also used more advanced engines and more potent fuels: kerosene and liquid oxygen (LOX) in American missiles, and kerosene and nitric acid in Soviet ones. The Soviet preference for nitric acid over LOX was a calculated risk, promoted by engineer Mikhail Yangel over the objections of his bosses Sergei Korolev and Valentin Glushko. It obliged engineers to design for, and missile crews to work with, a highly corrosive liquid, but it allowed missile fuels to be stored for long periods without the special cooling equipment and insulated tanks required to keep liquid oxygen liquid. Soviet missiles using kerosene and nitric acid could, therefore, be kept in high-alert condition—

---

2. The explosive power, or "yield," of a nuclear weapon is measured in thousands of tons (kilotons, abbreviated KT) or millions of tons (megatons, abbreviated MT) of the chemical explosive TNT. The largest conventional bomb used in World War II, called the "Grand Slam," contained 5 tons of chemical explosive. The bomb that destroyed Hiroshima had a yield of roughly 15 KT. The standard warhead used on the R-5M had a yield of 1 MT, and the Redstone's had a yield of 3.75 MT.

fully fueled and on the launch pad—for longer than American missiles. They also took less time to fire, since nitric acid (unlike LOX) could be pumped into the missile's tanks before the order to fire was received.

Improvements in engine design gave the second generation of ballistic missiles significantly longer ranges than the first: 1,200 miles for the Soviet R-12 (designated the SS-4 by the West); 1,400 and 1,600, respectively, for the American Thor and Jupiter; and over 2,000 miles for the Soviet R-14 (the SS-5). The new missiles' extended range meant that there was no need to deploy them close to where the front lines would be in the event of a war, or to make them "road mobile" (as the V-2 had been) so that they could be moved away from oncoming enemy forces. The new intermediate range ballistic missiles (IRBMs) could be deployed at permanent bases far from the targets they were intended to strike, yet still reach their targets in a matter of minutes. They gave the United States the power to hit targets in Soviet-controlled Eastern Europe from bases belonging to fellow members of the North Atlantic Treaty Organization (NATO). Simultaneously, they gave the Soviet Union the power to strike NATO countries (and China) from bases on its own soil.

IRBMs, unlike the short-range first-generation missiles, were "theater weapons." They had the capability to expand a war beyond the front lines and beyond the country in which the war had broken out. Using them could, in other words, turn a local conflict into a regional one. Any decision to use IRBMs, therefore, had to be as much a political choice as a military one. Decisions about where to base IRBMs also demanded a careful balancing of political and military factors. Misjudging the balance could be potentially disastrous, as the Cuban Missile Crisis of 1962 showed.

The deployment of the Jupiter to bases in Italy and Turkey in 1961 gave the United States a missile capable of hitting targets in the Soviet Union. Soviet R-12s and R-14s could hit targets in Western Europe (such as Thor bases in Britain) and China, but not the United States itself. Soviet premier Nikita Khrushchev found this imbalance galling, and sought to redress it by deploying several squadrons of R-12 and R-14s to Cuba in the fall of 1962. Militarily, this was sound thinking. It put the United States in the same strategic position that the Soviet Union had been in for a year: vulnerable, with nuclear-tipped IRBMs only minutes away from its major cities. Politically, it was a serious mistake. Cuba had become available to the Soviet Union only because communist dictator Fidel Castro had overthrown U.S.-backed dictator Fulgencio Batista in 1959. It remained available only because U.S. efforts to topple Castro—including the disastrous Bay of Pigs invasion in 1961—had failed. The presence of Soviet missiles in Cuba thus highlighted recent U.S. setbacks in the region. It also flouted the

160-year-old Monroe Doctrine, which declared the Western Hemisphere off-limits to the military and political adventures of European powers. Soviet missiles in Cuba were, in short, a challenge that no American president could allow to stand.

The discovery of the missile launch sites by U.S. reconnaissance planes triggered a tense, two-week diplomatic confrontation that brought the superpowers to the brink of war. President John F. Kennedy and his senior advisors contemplated a variety of options—including invasion and air attacks with conventional weapons—before eventually settling on a naval blockade and a demand that the missiles be removed. The crisis was eventually resolved by a secret agreement, the full details of which became public only after the fall of the USSR. Soviet premier Nikita Khrushchev publicly agreed to withdraw the R-12s and R-14s from Cuba immediately while Kennedy privately agreed to withdraw the Jupiters from Turkey and Italy the following year.

The resolution of the Cuban Missile Crisis ended a brief but important period in which IRBMs dominated superpower thinking about how to deter—or, if necessary, fight—a nuclear war. Even before the last Jupiters were pulled out of Turkey, however, the superpowers had better weapons with which to threaten each other.

ICBMS, SLBMS, AND NUCLEAR STRATEGY

Intercontinental ballistic missiles (ICBMs) and submarine-launched ballistic missiles (SLBMs) are exactly what their names imply. The former are weapons with ranges measured in multiple thousands of miles, capable of striking targets on the far side of the world. The latter are, essentially, IRBMs designed to be launched from submerged submarines. The nearly simultaneous introduction of ICBMs and SLBMs in the early 1960s meant that the superpowers could, for the first time, reliably strike each other's homeland with missiles fired from bases completely under their control.

The Soviet Union's first ICBM, the R-7, flew for the first time in 1957. Designed by Sergei Korolev, it had been conceived in 1950, but not authorized by the Soviet government until 1954. Built, tested, and brought to operational status in a crash program involving dozens of research institutes, the R-7 was—like the R-12 and R-14—a significant advance over earlier designs. It burned LOX and kerosene, rather than the LOX and alcohol of earlier Soviet missiles, and had two independent guidance systems: one inertial, one radio-controlled. It could carry payloads of 12,000 pounds over ranges of 3,000 miles—just enough, if launched from the right base, to hit targets in the United States with a thermonuclear warhead. An improved version, the R-7A, soon extended the range to nearly 5,000 miles. The R-7 family's most striking break with the past, however, was its layout. It consisted of a core stage surrounded by four booster stages, each of which tapered to a point and angled inward toward the core stage (see Figure 6.1). Each of the five sections contained four relatively small engines rather than one large one—a critical weight-saving measure. The sixteen engines in the four booster sections fired together at liftoff, to be joined shortly afterward by the four engines of the core section. A separate set of small, gimbaled vernier engines—two on each booster section and four on the core section—steered the rocket in response to signals from the guidance system.

The pace of ICBM development in the United States tracked the pace of development in the Soviet Union. The first American ICBM, named Atlas, was approved in 1951 as the Soviets tested the R-1 and R-2. Five years later, as the scope of Soviet missile programs became clear, work on Atlas was accelerated and development of a second ICBM, Titan, was begun. The Atlas became operational in September 1961 and the Titan in April 1962.

The Atlas, like the R-7, was a radical step forward in missile design. It was powered by three LOX-kerosene engines—two boosters and one sustainer—and used two smaller LOX-kerosene "vernier" engines to fine-tune thrust and steering (see Figure 6.2). The Atlas was designed to take off with all three main engines firing, shedding its boosters once they were exhausted and finishing the powered phase of its flight with the sustainer alone. The point of this "stage-and-a-half" design was to avoid having to ignite the sustainer in flight at high altitude—a major engineering challenge at the time. Atlas was equipped with the latest all-inertial guidance system and separable reentry vehicle, allowing it to reliably place its warheads within 600 yards of a target. Equipped with a large enough nuclear warhead—1 megaton or larger—it was capable of destroying even blast-resistant targets, such as missile silos. The most striking thing about the Atlas was its ultra-lightweight structure, which one engineer compared to an aluminum balloon pressurized by the fuel inside. Titan, a more conservative design, had two distinct stages—the first with two LOX-kerosene engines, the second with one—and separate tanks mounted inside a conventionally rigid, framed body. Its performance was comparable to that of the Atlas, with a slightly larger payload compensating for a somewhat shorter range.

The first ICBMs were designed (as IRBMs had been) to be stored horizontally and then raised to vertical for fueling and firing. The problem

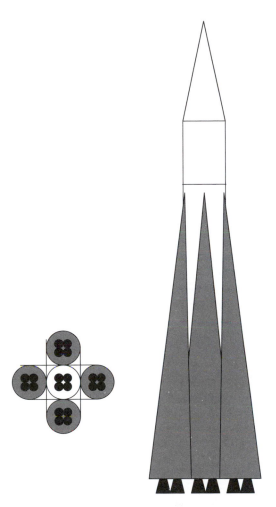

Figure 6.1: Schematic bottom view (left) and side view (right) of the R-7, showing the arrangement of the core stage (white), four booster stages (gray), and engine nozzles (black). Designed by Sergei Korolev, the R-7 was the USSR's first ICBM and the technological ancestor of the long-lived Soyuz family of launch vehicles. Drawn by the author.

with this system was that it made the missile vulnerable: whether in their storage sheds or on the pad, early ICBMs were likely to be destroyed if a nuclear bomb exploded nearby. A surprise enemy attack—a "first strike" or "nuclear Pearl Harbor"—could destroy a country's entire ICBM force on the ground, leaving it unable to retaliate. A country that believed it could carry out such an attack might, therefore, be encouraged to try it. A country

Figure 6.2: NASA employees watch as an Atlas missile is unloaded from a military transport in 1961. The two men standing near the tail of the missile give an idea of its size. Atlas, the first American ICBM, was a major technological step beyond earlier missiles and the first U.S. launch vehicle capable of putting a manned spacecraft in orbit. It launched John Glenn's three-orbit flight in February 1962, and the three Project Mercury flights that followed it. Courtesy of NASA Kennedy Space Center, image number GroundAtlas.

that believed itself vulnerable would be encouraged to launch its missiles at the first sign such an attack was imminent ("launch on warning"), rather than wait and risk losing its most powerful weapon. The vulnerability of early ICBMs thus encouraged a "use it or lose it" attitude that made war more likely.

The U.S. Air Force eventually solved this problem—as did the USSR's Strategic Rocket Forces—by basing its ICBMs in buried vertical tubes called "silos." Early missile silos were basically buried storage boxes with thick concrete walls and heavy steel lids that lay flush with the surface of the ground. The missile would sit inside, safe from anything but a direct hit by a nuclear bomb. If and when it had to be used, an elevator would lift it to the surface for fueling and final preparations. Titan was designed from the start to be stored in silos, and later versions of the Atlas were redesigned for silo basing. The Soviet Union developed silo-based versions of its R–12 and R–14 IRBMs in the early 1960s. The R–16, a Yangel design, using an R–14 for a first stage and an R–12 for a second stage, introduced another critical innovation: the ability to launch directly from the silo. The United States also developed its silo-launched ICBMs in the early 1960s. The Titan II, first flown in 1962 and first launched from a silo in 1963, was a major redesign of the well-tested Titan I. It still consisted of two conventional stages with a total of three engines between them, but now burned a storable combination of hydrazine as a fuel and nitrogen tetroxide as an oxidizer and could be launched from its silo on sixty seconds' notice. The Titan II far outperformed its namesake, extending the range by half (6,300 to 9,300 miles) and doubling the size of the warhead (4.5 MT to 9 MT). The Minuteman I, which first flew in 1961, used solid-propellant rocket motors, making it even easier to maintain and quicker to fire than the Titan II.

It was no coincidence that the first SLBMs entered service at roughly the same time as the first silo-launched ICBMs. They depended on many of the same technological breakthroughs, although for different reasons. Large solid-propellant motors were valuable in ICBMs because they could be stored for long periods of time and fired on a minute's notice. They were essential in SLBMs because volatile liquid propellants were too dangerous to store and handle in the tight, enclosed spaces aboard submarines. Silo launching was useful on land because it reduced the missile's vulnerability and shortened its reaction time. It was essential at sea because it saved critical space (no need for cranes or elevators) and eliminated the need to stand a missile upright on a rolling, pitching deck. Silo-launched ICBMs of the early 1960s were propelled out of their silos by their own exhaust gasses in what was called a "hot launch." Rather than find a way of igniting a rocket motor inside a ship, SLBM designers developed a different method, known as a "cold launch," that ejected the missile from its tube with a burst of compressed gas. The missile's own motors would ignite once it was clear of the ship.

The final technological breakthrough that made SLBMs a practical weapon was the development of the nuclear-powered submarine. Nuclear submarines, like the diesel-and-battery-powered submarines of World War II, could maneuver in three dimensions at will and submerge in order to escape pursuing enemies. They also, however, had a crucial advantage over the older diesel boats: with no need for air to run their engines, they could remain submerged for days at a time instead of hours, and at sea for months instead of weeks. They could, therefore, cruise the world's oceans at will, even passing under the Arctic ice cap. This freedom of travel made them difficult to find and impossible to target in advance. ICBM silos were fixed, and their positions were known. If sufficiently powerful, accurate missiles became available, they could be marked for destruction before war broke out. Missile submarines, and thus the SLBMs they carried, could be anywhere, waiting for an order to fire. This, along with the fact that early SLBMs had relatively small warheads and limited accuracy, made them ideal "second-strike" weapons: ineffective for surprise attacks against an enemy's missiles, but ideal for retaliatory attacks against an enemy's cities.

Deterrence is effective only if a nation can convince its enemies that any attack will result in the "assured destruction" of the attacker's homeland. ICBMs based in hardened underground silos and SLBMs based on hidden submarines made it easier to create and sustain that conviction. Their capacity for destruction and their near-invulnerability became central to the doctrine of "mutual assured destruction" (MAD). The basic concept of MAD was simple: No matter which side starts a nuclear war, no matter which side "wins" it, both sides will lose everything by the time it is over. Rational leaders, the theory went, would thus seek to avoid nuclear war at all costs.

The advent of SLBMs and silo-based ICBMs in the early 1960s made the doctrine of MAD frighteningly plausible. The premise of MAD was that, if each superpower believed that the other could absorb a full-scale nuclear attack and *still* be able to retaliate, neither would dare to attack the other. What rational leader would order an attack that would guarantee his own country's destruction? The new basing systems ensured that a substantial number of missiles would survive even the most vicious of first strikes. The new types of missiles ensured that the surviving missiles would be enough to deliver a killing low. Doing so, in the inside-out logic of the Cold War, made the world a safer place.

BALLISTIC MISSILES AND SPACE

The Cold War was not solely a military duel. The superpowers also engaged in symbolic competition: in science, in international sports, and especially in the exploration of space. Both the Soviet and the American space programs were nominally peaceful enterprises, but neither could have existed without ballistic missiles. Nor could the space around Earth have been so quickly or completely filled with satellites without modified ballistic missiles to put them there.

Sputnik I, the first artificial satellite in history, rode to orbit on October 4, 1957, atop an R-7 ICBM. The flights of *Sputnik I* and *Sputnik II* (launched a month later) were, in fact, part of the testing program for the new missile. Project Mercury, the first phase of the American manned space program, began in 1961 by using Redstone missiles to launch single-seat spacecraft on high-parabolic trajectories. It concluded, in 1962–1963, with four flights that used the more powerful Atlas missile to place similar spacecraft in orbit. Project Gemini—the second phase of the program, consisting of ten missions flown in 1965–1966—used modified Titan II missiles to lift a larger, heavier spacecraft capable of carrying two astronauts for fourteen days.

Despite these high-profile successes, however, the use of unmodified or slightly modified missiles as launch vehicles virtually ceased by the mid-1960s and was uncommon even before then. Military missiles met the basic requirements for a space launch vehicle: they could supply massive amounts of thrust, and so could carry significant payloads for substantial distances. Delivering a satellite or spacecraft to a specified orbit is a different problem, however, than delivering a nuclear warhead to a specified target. A missile warhead is meant to be pulled back to Earth's surface soon after launch; a satellite or spacecraft, on the other hand, succeeds only if it achieves sufficient altitude and velocity to remain in orbit.[3] The heavier the payload, the higher the orbit, or the greater the angle between the plane of the orbit and the plane of the Earth's equator, the more power is required. Configured as an ICBM, the Atlas E missile could deliver a 5,500-pound warhead to a target 7,500 miles away; configured as a launch vehicle, the same missile could lift an 1,800-pound payload to an orbit 110 miles above the Earth's surface.

---

3. A projectile on Earth (a missile's warhead, for example) is carried forward by inertia and pulled downward by gravity. A satellite or spacecraft in orbit around the Earth is affected by the same two forces, but is moving forward so fast that it falls "around" the Earth rather than toward its surface.

Performance like that was not sufficient for the kinds of missions that Soviet and American space program officials envisioned. They began, therefore, to develop ways to give "stock" military missiles more power.

One approach, of course, was to redesign the missiles themselves: making their bodies lighter, their fuel tanks bigger, or their engines more powerful. A second was to attach self-contained "strap-on" boosters that would work in concert with the missile's main engines. A third approach was to add one or more upper stages to the basic missile. Soviet designers, who had already begun to use these concepts in ICBM design, quickly applied them to launch vehicles as well. The Vostok launch vehicle that made Yuri Gagarin the first human in space on April 12, 1961, was an R-7 missile with a small second stage added. Swapping the Vostok's second stage for a larger one created the more powerful Voskhod. A series of still-larger second stages created the Soyuz (still in use at this writing) and the addition of a third stage produced the Molniya family of satellite launchers. American designers applied the same approach to the Atlas and Titan, which by the mid-1960s were beginning to be retired in favor of newer Minuteman ICBMs. Outfitted with strap-on boosters and upper stages, however, the Atlas and Titan gave decades of service as launch vehicles and remain in service at this writing.

The most successful family of launch vehicles in U.S. service, however, was based on an undistinguished Air Force IRBM named Thor. Even before its brief deployment as a weapon in the early 1960s, the Thor was being modified for other duties. Fitted with a liquid-propellant upper stage and dubbed Thor-Able ("Able" being the name of the upper stage), it was used to test reentry vehicles for the then-forthcoming Atlas. Substituting a larger, more powerful "Agena" upper stage enabled it to boost the "Corona" spy satellites of the late 1950s into polar orbits that carried them over the Soviet Union. A Thor-Able III launch vehicle—the Thor first stage, plus an improved Able upper stage—put the scientific satellite *Explorer 6* into orbit in 1959, where it returned the first television images of the Earth from space. Able broke new ground in using "hypergolic" propellants: liquids that would spontaneously ignite on contact, even in the vacuum of space. Increasingly involved with launches of military satellites, the Air Force continued to develop Thor-Able and Thor-Agena launch vehicles (see Figure 6.3).

The same year that *Explorer 6* flew, NASA announced plans to develop a new three-stage launch vehicle by adding a small, solid-propellant third stage to the basic Thor-Able configuration. Delta formally entered NASA service in 1960, carrying a series of important scientific and communications satellites into Earth's orbit. It was followed in 1962 by the Delta A

Figure 6.3: The Transit IV-A satellite, powered by an onboard nuclear reactor, is launched from Cape Canaveral aboard a Thor-Able launch vehicle on June 29, 1961. The joint between the Thor first stage and the Able upper stage is just above the letter U in "USAF." Atomic Energy Commission photograph. Courtesy of the National Archives at College Park, Maryland, NAIL image number NWDNS-326-PV-(4)185(1).

(which had a more powerful first-stage engine) and Delta B (larger second-stage tanks), and in 1964 by the Delta C (an improved third stage). The Delta D, introduced in 1964, added three strap-on, solid-fuel boosters to the first stage. After an across-the-board engine upgrade to produce the Delta E in 1965, the design remained relatively stable until the M-6 and N-6 models introduced at the end of the decade, which featured a larger first-stage fuel tank and six strap-on boosters instead of three. The current version of Delta was rushed into service in the late 1980s after the destruction of the *Challenger*, and the grounding of the space shuttle fleet created a backlog of payloads. Named Delta II, it uses an even longer first-stage fuel tank, up to nine strap-on boosters, and an enlarged payload fairing that gives it a distinctive, bulbous nose (see Figure 6.4). Still in service at this writing,[4] it is still recognizable as a direct (if distant) descendant of the long-retired Thor.

Launch vehicles derived from ballistic missiles gave the industrialized world ready access to orbit around the Earth. They enabled humans to enter outer space for the first time, set the stage for the "moon race" of the 1960s, and launched the robot spacecraft that (between 1962 and 1980) delivered the first close-up observations of Mercury, Mars, Venus, Jupiter, and Saturn. Most important, they made it possible to put satellites into orbit around the Earth quickly, reliably, and (relatively) economically.

Satellites returned pictures of things that were invisible, or nearly so, by conventional means. They documented troop movements, nuclear weapons tests, and construction of new missile silos. They transformed the science of weather forecasting, and dramatically improved meteorologists' ability to track hurricanes, predict their paths, and issue evacuation warnings to low-lying areas. Satellite photography lent support to the emerging environmental movement of the 1960s by documenting the scale of ocean dumping, deforestation, and other human activities. Satellites' ability to relay electronic signals over vast distances revolutionized communications, and quickened the pace at which diplomacy, journalism, and international trade moved. Satellites freed international telephone communications from their long dependence on expensive and sometimes unreliable transoceanic cables. They gave political and military leaders unprecedented ability to understand and control crises as they unfolded. They allowed television viewers to experience, in real time, events taking place on the far side of the world: a variety show in London (1962), the Olympics in Tokyo (1964), and so on. Many of the most famous images of late twentieth and early

---

4. The two *Mars Odyssey* spacecraft that landed in early 2004, for example, were launched aboard Delta IIs in mid-2001.

Figure 6.4: A Delta launch vehicle, augmented by strap-on boosters and carrying the Nimbus 5 weather satellite, waits on its launch pad at Cape Canaveral in 1972. The Delta, which evolved from the Thor-Able pictured in Figure 6.3, has been the most successful of all U.S. launch vehicles. An updated version, the Delta II, is still in service. Courtesy of NASA Headquarters, image number Nimbus-Delta.

twenty-first centuries—the fall of the Berlin Wall, the Tiananmen Square protests in Beijing, and the collapse of the World Trade Center towers—are universally recognized because satellite communications made them instantly, widely available.

THE ARMS RACE AND ARMS CONTROL

Each superpower had, by 1960, deployed missiles capable of striking the other's homeland with nuclear warheads. Each sought, over the next decade, to ensure that they possessed a "credible deterrent": a nuclear strike force capable of surviving an enemy attack and carrying out a devastating counterattack. Missiles were central to that goal, and both superpowers expended vast amounts of time and treasure to acquire them. Soviet and American leaders alike worried about the number of missiles in their arsenals. John F. Kennedy, in his 1960 campaign for the presidency, attacked the Eisenhower administration for allowing a "missile gap" to develop between the United States and the USSR. Sheer numbers, however, were not the whole story. Rapid improvements in ballistic missile technology meant that the capabilities of particular missiles mattered just as much.

The Soviet R-36 ICBM (known as the SS-9 "Scarp" in the West) set off alarm bells in Western intelligence agencies when it was first tested in 1963. Reports suggested that it would be both more accurate and more powerful than existing Soviet ICBMs, with a CEP of only a mile and enough payload to carry a 25-MT nuclear warhead. That combination of accuracy and power made the R-36 a serious threat to Minuteman missile silos and, even more critically, to the underground launch control centers (LCCs) in which their crews worked. Each LCC controlled ten silos, so the loss of even a single one could be devastating. American military analysts worried that the SS-9 might be designed as a "first-strike" weapon, capable of demolishing American ICBMs in their silos during the first moments of a war. Responding to the perceived threat of the R-36, the United States built more Minuteman silos and networked the LCCs so that, if one was destroyed, other crews could take control of its missiles.

The United States created similar anxiety in the Soviet Union by developing and planning to deploy "antiballistic missiles": defensive weapons designed to destroy incoming missiles or reentry vehicles before they reached their targets. Ideas for antiballistic missile (ABM) systems had been around since the mid-1950s. The technology that would make them practical— small, high-performance rocket engines and miniaturized electronic components for radars and guidance-system computers—took another dozen years to arrive. By 1969, however, newly elected President Richard Nixon was impressed enough to approve plans for a twelve-site ABM network to be called Safeguard. The principal mission of Safeguard was to protect ICBM silos against Soviet missiles: to blunt a full-scale attack or destroy a missile launched by accident. The Soviets were also at work on an ABM system, but lagged well behind the Americans. They saw Safeguard, therefore, as a threat to the power of their nuclear forces. The greater the percentage of Soviet missiles the United States could shoot down, the less they would fear a Soviet nuclear attack.

The development of multiple independently targeted reentry vehicle (MIRV) technology in the 1960s affected both superpowers. Early ICBMs and SLBMs had carried only a single reentry vehicle and so could hit only a single target. Missiles equipped with MIRVs could carry multiple warheads (originally three, later more), each of which could be programmed to hit a separate target. MIRVs thus multiplied the destructive potential of a missile. The United States hoped, in the late 1960s, to use their lead in MIRV technology to offset the effects of the Soviet Union's massive missile-building program. Having fewer missiles than the enemy was tolerable, the argument went, if some of your missiles could carry more than one warhead. Despite the short-term advantage it offered the United States in the late 1960s, MIRV technology posed problems in the long term. The prospect of the Soviet Union continuing to build more ICBMs, while at the same time adding MIRVs to them, suggested that the arms race would only accelerate in the 1970s.

The Strategic Arms Limitation Treaties of 1972 and 1979, known informally as SALT I and SALT II, were designed to reign in the arms race. Negotiations for SALT I began in Helsinki, Finland, in the fall of 1969. Each superpower hoped to use SALT to magnify its own advantages in missile technology and limit its opponents' advantages. Soviet leaders wanted to limit the talks to defensive weapons, enabling them to stop deployment of ABMs while using MIRVs to multiply their advantage in ICBMs. American leaders wanted the talks to include both offensive and defensive weapons, enabling them to preserve the existing advantage in MIRVs that gave the United States as many warheads as the Soviet Union despite having fewer missiles. The talks, predictably, bogged down. They remained bogged down until May 1971, when a flurry of diplomatic activity broke the deadlock and resulted in two historic agreements signed by President Richard Nixon and Soviet premier Leonid Brezhnev on May 26, 1972. The first agreement, the ABM Treaty, limited each country to two ABM sites with 100 missiles each: one site to protect the capital city and one to protect an ICBM field. The second, the Interim Agreement, froze the superpowers' missile arsenals at their existing levels for five years. These limits preserved the Soviet advantage in total numbers of missiles (2,328 ICBMs and SLBMs compared to 1,710), but also the American advantage in MIRVs. Equally important, the two treaties set a powerful precedent: that the two superpowers were willing to voluntarily limit their arsenals.

The SALT II talks began in Geneva in mid-1973 and, like the SALT I talks, bogged down almost immediately. They surged forward, however, during a November 1974 summit in Vladivostok, when President Gerald Ford and Soviet leader Brezhnev agreed in principle to limit their arsenals to 2,400 long-range missiles, no more than 1,320 of which could carry MIRVs and no more than 330 of which (on the Soviet side) could be "heavy ICBMs" like the R-36. The tentative agreement bogged down a second time, however, when Ford tried to win Congressional support for it. Debate and renegotiation dragged on until 1979, when the Senate killed the treaty by refusing to ratify it. Many observers loudly declared arms control dead.

An ongoing Soviet missile buildup lent credibility to that view. The SS-17 and SS-18 ICBMs entered service in the mid-1970s, as did the SS-20 IRBM. All three belonged to a new generation of Soviet missiles, and each had a CEP of 400 meters or less—three times more accurate than the missiles they replaced. The SS-17 was the first operational Soviet ICBM to carry MIRVs, and the first to use "cold launch" technology. The SS-18, code-named "Satan" by NATO, was designed to replace the old SS-9. It was accurate enough, and its warheads—ten 500-KT MIRVs or one massive 20-MT device—were powerful enough to destroy even hardened American ICBM silos. The SS-20, a two-stage missile designed to be moved by road and fired from temporary launch sites, was officially an IRBM aimed at targets in Western Europe. The addition of a third stage, however, gave it sufficient range to reach the United States.

Conservative politicians in the United States watched these developments with growing dismay. Ronald Reagan campaigned for—and won—the presidency in 1980 partly by focusing on what he called a "window of vulnerability" that put the United States at risk. The Soviet Union could devastate NATO with its SS-20s, Reagan and his supporters argued, or unleash a surprise first strike on the United States that would wipe out up to 80 percent of the American ICBM force in their silos and still leave 1,000 ICBM warheads in reserve. The Soviets could also, conservatives feared, compel NATO's surrender in a conventional European war simply by threatening to unleash their missiles. Defense policy in the first Reagan administration centered on a new generation of American missiles: Peacekeeper ICBMs and Trident II SLBMs to deter an attack on the United States, plus Pershing II IRBMs and ground-launched cruise missiles (GLCMs, pronounced "glick-ems") to protect Europe by threatening targets in the Western USSR.

The technological centerpiece of the first Reagan administration, however, was not a missile but a space-based missile-defense system. The system, as originally conceived, was to be a network of satellites armed with lasers and other futuristic weapons and controlled by computers that would enable them to track and destroy hundreds of Soviet ICBMs as they arced through the atmosphere. Called the Strategic Defense Initiative by administration officials, it was quickly dubbed "Star Wars" by the public and press. Soviet leaders protested that it would violate the ABM treaty and allow the United States to attack the USSR with impunity. Many scientists protested that the system depended on technologies that had scarcely been invented, and could never be fully tested unless and until it had to be used. The criticisms, however, had little impact on Reagan. His deeply held conviction that the United States could be made safe from nuclear missiles forever sustained the project until it quietly expired along with the Cold War itself.

The direction of the arms race changed—suddenly, rapidly, and unexpectedly—beginning in 1985. Reagan developed an improbably good working relationship with the new leader of the Soviet leader, reform-minded Mikhail Gorbachev. Together at a summit in Reykjavik, Iceland, on October 11–12, 1986, they agreed informally to cut their strategic nuclear forces in half and came within (Reagan later said) "one lousy word" of agreeing to abolish their nuclear arsenals altogether. Neither proposal reached the treaty stage, but they set the stage for strategic arms reduction agreements signed by Reagan's and Gorbachev's successors in the 1990s. The summit did lead, however, to the Intermediate-Range Nuclear Forces (INF) treaty, signed in December 1987. In it, the United States and USSR agreed to destroy all nuclear missiles with ranges between 300 and 3,400 miles: 859 American GLCMs and Pershing IIs, and 1,836 Soviet SS-20s. It was the first treaty in history to reduce (rather than limit) nuclear arsenals, and the first to eliminate an entire class of missiles.

MISSILES IN OTHER COUNTRIES

The superpowers were not the only countries to see ballistic missiles as essential tools for deterring their enemies and extending their military power beyond their borders. The United States supplied weapons, including short- and intermediate-range ballistic missiles, to fellow members of NATO, and the Soviet Union did the same to its allies: the signers of the Warsaw Pact. Both superpowers also supplied missiles to allies in the Middle East, where the U.S.-backed state of Israel fought a series of wars with its Soviet-backed Arab neighbors. France, Britain, Israel, and China pursued ballistic missile programs of their own between the late 1950s and the late 1970s, in order to achieve a degree of technological independence.

Most of the ballistic missiles that the superpowers supplied to their European allies were mobile systems capable of carrying small nuclear warheads. The U.S.-built, truck-mounted Honest John, widely used by NATO armies in the late 1950s and 1960s, could hit targets 20 miles away with a 20-kiloton warhead. The Lance, which replaced it in the early 1970s, could carry warheads ranging from 1 to 100 KT over ranges up to 60 miles. The Soviet Union deployed a series of similar truck-mounted systems known by the acronym FROG ("free rocket over ground"). The FROG-7, which entered service in 1965, was not strictly a ballistic missile. Lacking a guidance system, it was really an advanced version of the *katyushas* of World War II. Nevertheless, it had performance comparable to that of Honest John and Lance: a range of 40 miles and the ability to deliver a 1,000-pound high explosive or chemical warhead or a 25-KT nuclear warhead. Warsaw Pact countries also received substantial numbers of R-1 missiles, dubbed the SS-1 "Scud" by NATO. Little more advanced than the V-2, the first versions of the Scud had ranges well under 100 miles and simple high-explosive warheads. Later versions—while still notoriously inaccurate—could reach over 400 miles and carry chemical or nuclear warheads in place of high-explosive ones.

The superpowers deployed these missiles less to prevent a war in Europe than to fight such a war if and when it broke out. Senior NATO commanders saw them as a means to neutralize the Soviets' advantage in conventional weapons. Senior Warsaw Pact commanders—who did not have direct control of nuclear warheads, but knew that their Soviet superiors could order them brought to the front and used—saw their own missiles as a threat that would make NATO think twice before crossing the nuclear threshold.

Soviet leaders, from the late 1950s onward, saw the Middle East as a politically important region where it was vital to cultivate allies and block Western expansion. They became, as a result, the region's biggest arms supplier, shipping large quantities of weapons to "client states" such as Egypt, Syria, and Yemen. Having armed the military forces of Syria and Egypt, the Soviet Union rearmed them with better weapons after their defeat by Israel in 1967, did so again after their defeat in 1973, and rearmed Syria a third time after Israel overwhelmed Syrian forces in Lebanon in 1982. These infusions of arms included shipments of FROG and Scud missiles. Egypt, for example, received the FROG-7 in 1968, Syria in 1973, and Algeria later in the decade. Egypt received the first Scud-B missiles to be deployed outside of Europe—a dozen launchers and over hundred missiles—in 1973. Over the next decade the Soviets spread Scuds throughout the region: Libya and Iraq in 1974, Algeria a few years later, and Yemen a few years after that.

They would continue to do so well into the 1980s, setting the stage for the use of missiles in the Iran-Iraq War and the first Persian Gulf War.

Israel, whose unique diplomatic relationship with the United States began with its founding in 1948, originally relied on France as its principal arms supplier. In the late 1960s, however, the United States took over the role in an effort to strengthen Israel against its Soviet-supplied Arab neighbors. American leaders shied away, however, from equipping Israel (or other Middle Eastern countries) with even short-range ballistic missiles. A 1973 shipment of a dozen Lance launchers and 150–200 missiles to Israel was the sole Cold War–era exception to this pattern. Israel's 1973 request for the longer ranged Pershing missile was turned down, and the shah of Iran— whose close ties to the United States eventually cost him his throne in 1979—received no ballistic missiles at all. Outside the highly volatile Middle East the United States was less restrained. It equipped the South Korean army with Honest John missiles in the late 1950s, and sent more to Taiwan in the early 1960s. It also supplied sounding rockets and launch vehicles— potentially the foundation for a ballistic missile program—to Argentina, Brazil, and India.

The handful of nations that developed their own ballistic missile programs in the 1960s and 1970s did so as gestures of independence. On one hand, the independence they sought was economic: a country that can manufacture its own weapons can reject a would-be patron's latest arms deal without walking away empty-handed. On the other hand, it was also political: a country that can build its own missiles is less likely to require outside help to defend itself, deter aggression, or impose its will on others.

Britain's IRBM program began in the late 1950s but collapsed in 1960 when the Blue Streak missile failed to live up to expectations. The British government fell back on its long-standing "special relationship" with the United States, accepting shipments of Thor IRBMs to be deployed at British bases and Polaris SLBMs to be deployed on British submarines. France, more determined than Britain to demonstrate its independence from NATO and the United States, also had better luck with its missile program. It developed the SR–3 IRBM in the 1960s, and deployed them in silos between Lyon and Marseilles in the early 1970s. France simultaneously supported Israel's ballistic missile program and helped to build and test prototypes of what became the Jericho missile. A 1968 arms embargo ended French involvement, however, and (as noted above) the United States was opposed to spreading ballistic missile technology in the region. Israel chose, therefore, to press on alone, spending over $1 billion on the Jericho program in the 1970s alone. Ultimately, however, Israel's efforts paid off. Once deployed in the late 1980s, the Jericho 1 gave Israel a missile capable of carrying a 1,000-pound payload (potentially, a small nuclear warhead) over 250 miles.

China acquired its first ballistic missiles—R-1s and a single R-2—from the Soviet Union in the late 1950s. They might have continued to use Soviet missiles had the two countries not broken off relations in 1960. Having acquired technical drawings and a single example of the Soviet R-12 missile before the rift, the Chinese set about developing their own missiles based on Soviet designs. The first of them, called the Dong Feng ("East Wind") 1, was a straightforward copy of the Soviet R-2, which was itself only slightly different than the German V-2. The Dong Feng 2, China's first IRBM, copied the design of the R-12 down to the use of storable liquid propellants. It could carry a nuclear warhead just over 600 miles—a modest range by Western standards, but enough to hit Japan or South Korea. Plans approved in 1964 called for a Dong Feng 3 capable of hitting the Philippines, a Dong Feng 4 capable of reaching the United States Air Force base on Guam, and a Dong Feng 5 that could hit the Western United States. All three missiles flew by the early 1970s, and all three were deployed (the DF-4 and DF-5 only in very limited numbers) by the early 1980s. China thus acquired, in the space of two decades, a full range of ballistic missiles.

The Chinese also modified their missiles, as the United States and the Soviet Union had, to launch payloads into Earth's orbit. Just as the Thor IRBM had provided the first stage for the Delta launch vehicle, so the DF-3 IRBM provided the first stage for the Changzheng ("Long March") 1 launch vehicle that carried China's first satellite to orbit in April 1970. The CZ-1 gave way, in the mid-1970s, to the CZ-2, based on the DF-5 ICBM, and it was from the CZ-2 that all subsequent launch vehicles in the Long March family were derived. Like many ICBMs—Atlas, Titan, and R-7 among them—Dong Feng 5s have spent far more time carrying peaceful cargoes to orbit than waiting to carry nuclear warheads to an enemy's homeland.

7

Rockets to the Moon, 1960–1975

◆

Addressing a joint session of Congress on May 25, 1961, President John F. Kennedy declared: "I believe that this nation should commit itself to the goal, before this decade is out, of landing a man on the moon and returning him safely to the Earth." It was a bold, visionary statement, but also a carefully calculated one.

Using a rocket to reach the moon was not a new idea in 1961. Tsiolkovsky, Goddard, and Oberth had all discussed it in print. Wernher von Braun, even as he built V-2s for the German army, had his eyes fixed firmly on the moon. The first "modern" science fiction film, *Destination Moon* (1950), told the story of a private lunar expedition aboard a futuristic nuclear-powered rocket. By the time Kennedy spoke to Congress in 1961, imaginary trips to the moon had become commonplace in popular culture. Americans had read about them in weekly magazines like *Collier's* and the *Saturday Evening Post*, seen them dramatized in movies (*Destination Moon*) and television (*Men into Space*), and experienced them firsthand in the "Trip to the Moon" simulator at Disneyland. NASA planning for manned flights to the moon began in the spring of 1959, and NASA chief Keith Glennan approved them on January 7, 1960. A week later, President Dwight Eisenhower authorized the "accelerated" development of a new "super booster" named Saturn. NASA engineers had, by the end of 1960, sketched out a rough schedule: flights around the moon in 1967–1968, and landings in 1969–1970.

Kennedy knew about NASA's plans when he stood before Congress in May 1961. He had been looking for a space "first" that the United States could achieve before the Soviet Union, and asked his advisors to confer with NASA officials and explore the possibilities. NASA reported that a manned lunar landing represented the best hope for an American victory. Vice President Lyndon Johnson backed the idea, promoting it at the White House and simultaneously pushing NASA to develop a concrete proposal that the president could review before his address to Congress. Kennedy's May 25 speech was, therefore, a call not to make plans but to act on plans already made. It was bold not because it offered a new vision of the future, but because it committed the nation's time, treasure, and technical expertise to a familiar one. It was bold, too, because it set a deadline: not "someday" or even "soon" but in less than ten years. Setting that deadline and staking the nation's prestige on meeting it gave the program instant, powerful, lasting momentum. It encouraged NASA officials to seriously consider bold innovations they might otherwise have rejected, led Congress to appropriate funds they might otherwise have held back, and interested the public in events they might otherwise have ignored.

THE PLAN

"Landing a man on the moon and returning him safely to the Earth" meant solving two enormous technological problems. One was designing the spacecraft itself: a vehicle capable of landing on a barely explored world a quarter-million miles from Earth and then taking off, hours or days later, to begin the quarter-million-mile journey home. The other was designing a booster capable of lifting such a spacecraft, its crew, and all their equipment off the Earth. Neither problem could be solved independently. The spacecraft had to fit onto, or into, the booster that would carry it. Its weight (including fuel, food, water, and other "consumables") had to be comfortably less than the booster's maximum payload. The booster, meanwhile, had to be big enough and powerful enough not to force the spacecraft's designers into weight-saving compromises that might endanger the mission or imperil the crew. The first step in designing either the spacecraft or the booster was to decide, in general terms, how the two would work together. American and Soviet engineers alike had three basic "modes" to choose from: direct ascent, Earth-orbit rendezvous (EOR), or lunar-orbit rendezvous.

Direct ascent would use a spacecraft assembled on Earth and designed to remain in one piece for the entire mission. Once free of Earth's gravity, the spacecraft would fly to the moon, land, take off again, and fly back to

Earth. Direct ascent allowed the spacecraft to be prepared, tested, and fully loaded while still on Earth, an advantage that appealed to spacecraft designers like Max Faget. Lifting such a spacecraft would take an immense booster, but Wernher von Braun's team had one on their drawing boards: a monster called Nova, planned to be twice as powerful as the biggest Saturn.

Earth-orbit rendezvous required a similar spacecraft, but not the lifting power of a Nova-class booster. One popular variation called for launching the spacecraft's crew cabin on one Saturn and the propulsion system on another, then joining the two in orbit. Another variation called for launching a fully assembled spacecraft with no fuel aboard, then fueling it from a separately launched tanker. Von Braun and his team became strong supporters of Earth-orbit rendezvous. They believed in Nova as a concept, but first wanted experience that would come from building and flying the Saturn. Earth-orbit rendezvous missions would give them that experience, and lay the foundation for later missions where Nova would be essential. The problem with the method was not building the hardware but making it work. Space walks, precision maneuvering in space, and rendezvous between spacecraft had never been tried in 1961. Earth-orbit rendezvous would take all those skills, and others that would be even more difficult to master.

Faget's spacecraft-design team had, by early 1962, uncovered a serious problem with both direct ascent and EOR. A spacecraft capable of going to the moon and back would be dangerously awkward to actually *land* on the moon. The ship (which might be as long as 90 feet) would have to descend tail first toward the lunar surface, slowing its descent with the same engine that had propelled it from Earth. The pilot, seated in the nose of the ship with his face toward the sky and his back parallel to the ground, would have to land virtually blind. Radar or backward-facing television cameras would be his only way to "see" the terrain beneath his landing gear. Soft or uneven terrain—even an unseen boulder underneath one landing leg—could topple the spacecraft onto its side, leaving it unable to ever lift off again. Walt Williams, a senior NASA engineer, compared such a landing to "backing an Atlas [missile] down onto the pad" (Murray and Cox 1989, 112) and shuddered at the idea of asking anyone to do it.

The third method of going to the moon, lunar-orbit rendezvous, used a spacecraft divided into two specialized parts: one for getting to and from the moon, the other for landing on it (see Figure 7.1). Carried into Earth orbit by a single Saturn booster, the two parts would be joined, nose-to-nose, for the flight to the moon. The streamlined, relatively spacious Command-Service Module (CSM) would carry a three-member crew, their equipment, and enough consumables to sustain them for fourteen days. It would also, with its powerful engine and large fuel supply, provide propulsion for both ships on the outbound leg of the trip. The angular, buglike Lunar Module (LM) would be activated only after the spacecraft was in orbit around the moon. The LM would then carry two astronauts to the lunar surface, serve as their base of operations while they explored, and bring them back to the

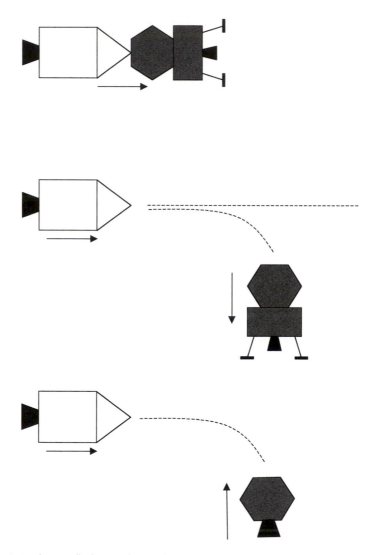

Figure 7.1: The Apollo lunar orbit rendezvous concept. Top: The CSM (white) and LM (gray) are joined nose to nose for the 3-day trip to the moon. Center: After achieving lunar orbit, the two spacecraft separate. The LM descends to the lunar surface using the engine in its lower ("descent") stage; the CSM remains in orbit. Bottom: The upper ("ascent") stage of the LM lifts off from the moon and makes its rendezvous with the CSM in lunar orbit. Drawn by the author.

waiting CSM orbiting above. The two ships would remain docked long enough for the LM crew to transfer themselves and their samples to the CSM. Then the LM, its job done, would be cast loose and left behind while the CSM took the astronauts back to Earth.

Lunar-orbit rendezvous had important advantages. The LM would have only one job, and could be designed specifically *for* that job. It would need neither streamlining nor a heat shield, because it would never fly in Earth's atmosphere. It would need to carry only enough consumables to support two men for three days. Above all, because it would make only two brief flights in the moon's low gravity, it could be small, light, and compact. It would, as a result, be far easier to land than a ship designed for a round trip from Earth to the moon. Lunar-orbit rendezvous also, however, had a serious drawback: the CSM was not capable of landing on the moon, and the LM not capable of returning to the Earth. If the postlanding ren-dezvous failed, the two astronauts aboard the LM would be stranded on the moon with no hope of rescue.

NASA engineers and mission planners were initially skeptical of the lunar-orbit rendezvous method, but it gradually gained support. John Houbolt, an engineer who championed it when few others were willing to do so, argued at length for its advantages. It would use the Saturn instead of the Nova, require one launch instead of two or more, and eliminate the need for low-gravity construction and fueling operations. These features also gave LOR a further advantage: its relative simplicity made it NASA's best hope for meeting President Kennedy's end-of-the-decade deadline. Engineers at NASA's newly established Manned Space Flight Center in Houston began to champion lunar-orbit rendezvous in early 1962. Both Faget and von Braun endorsed the plan by late spring, and NASA officially adopted it in early July.

THE HARDWARE: SATURN AND APOLLO

Lunar-orbit rendezvous, though less technologically demanding than either alternative, still posed huge challenges. It meant creating a booster of un-precedented power and a spacecraft of unprecedented complexity. The new booster (Saturn) would require ground facilities beyond anything that had been built before. The new spacecraft (Apollo) would require a more sophisticated series of flight tests than any that had been attempted before. Neither vehicle—for reasons both practical and political—would be built by a single company. The first stage of the booster would be built by Boe-ing, the second (along with the engines for all three) by North American, and the third by McDonnell-Douglas. The CSM and LM would be built a continent apart, one by North American in Seal Beach, California, and the other by Grumman in Bethpage, New York. The booster guidance system would come from IBM, the emergency escape system from Lockheed, and the parachute system from Northrop. Well before the end of the decade, all the components would have to come together, fit perfectly, and function smoothly.

The Saturn family of boosters had a lineage stretching back to the Thor and Jupiter rockets of the 1950s and, still further back, to the Redstone and the V-2. The Saturn I, designed in 1958 and first launched in 1961, used the same H-1 engines that had powered the Thor and Jupiter.[1] Eight of them, grouped in the first stage, gave it 1.5 million pounds of thrust at liftoff. The Saturn IB, first launched in 1966, had a similar first stage—eight H-1 engines burning kerosene and liquid oxygen—topped with a new second stage powered by a single, high-efficiency J-2 engine burning liquid hydrogen and liquid oxygen. The smaller Saturns were the first American boosters designed specifically to carry heavy payloads to Earth orbit, and the first that had not started life as missiles. They performed steadily throughout the 1960s and into the 1970s in vital but unspectacular supporting roles. The Saturn V, however, was built with the moon in mind. It overshadowed the Saturn I and IB even before it flew.

The Saturn was the ultimate expression of an approach to rocket design that began with the V-2. It was a sleek, multistage, liquid-fueled machine taller than the Statue of Liberty and heavier than a U.S. Navy destroyer (see Figure 7.2). Everything about the Saturn V was big. Each of the five F-1 engines in its first stage could produce 1.5 million pounds of thrust—as much as an entire Saturn I first stage. All together, the three stages of a Saturn V produced as much thrust as 100 Redstones. Engineers calculated that, if a Saturn exploded at the moment of launch, the blast would equal that of a fair-sized nuclear bomb. The Saturn V's flight plan was an elegant demonstration, on a huge scale, of the centuries-old idea of staged rockets. The first stage, fed kerosene and liquid oxygen by pumps the size of small cars, would lift the Saturn-Apollo "stack" to a height of 36 miles and accelerate it to nearly 6,000 mph. The second stage, powered by five J-2 engines burning liquid hydrogen and liquid oxygen, would carry the third stage and spacecraft nearly to orbit: 108 miles and 17,500 mph. The third stage, driven by a single J-2, would complete the trip to orbit and then—in a second, separate, longer "burn"—give the Apollo spacecraft its initial push toward the moon.

Apollo was an even bigger step beyond earlier spacecraft than Saturn V was beyond earlier boosters. The cone-shaped command module was

---

1. The Saturn series used Roman numerals in their names, so "Saturn I" and "Saturn V" are pronounced "Saturn one" and "Saturn five."

Figure 7.2: A complete Saturn-Apollo "stack," photographed from the upper floors of the Vehicle Assembly in 1967. This particular vehicle is AS-501, which was used for the Apollo 4 mission (the first all-up test of the Saturn launch vehicle and Apollo spacecraft) in November 1967. Courtesy of NASA Kennedy Space Center, image number 67P-0208.

a classic Max Faget design, cut from the same cloth as Mercury and Gemini. The large, cylindrical service module, which rode behind it until just before reentry, was something else altogether. The "Space Propulsion System" engine in its tail, designed to soak for days in the cold of space and still fire on cue, would supply the thrust necessary to go to the moon and back. The equipment filling its storage bays—tanks, pumps, fuel cells, wires, and pipes—would supply much of the air, water, and power needed to keep the crew alive. The command module by itself could still be called a "capsule," as the Mercury and Gemini spacecraft had often been. The addition of the service module made it a true spaceship: a fully independent vessel capable of long, independent voyages.

The lunar module was the greatest technological leap of all, for the simple reason that no one had ever tried to build such a machine before. There was no successful design to study, and no fund of experience to draw on. The engineers at Grumman had to start, literally, with a blank sheet of paper and a list of requirements from NASA. The requirements were daunting. The LM had to weigh less than 30,000 pounds and fit inside a special "spacecraft adapter" that would join the Saturn's third stage to the Command-Service Module. It had to carry two astronauts and all the food, water, air, and equipment they would need for up to three days on the moon. It had to land safely on virtually unknown terrain *and* be able to lift off again safely several days later. Early in the design process, the LM became a two-stage design. The ascent stage, a lopsided collection of angles and bulges, sat atop the four-legged, octagonal descent stage (see Figure 7.3). The descent-stage engine would brake the LM as it descended toward the lunar surface. A second engine would lift the ascent stage, with two astronauts inside, back to its orbital rendezvous with the CSM. The two-stage system solved many problems, but added a new one: both engines had to work flawlessly even after the stresses of liftoff and days in the cold of space. Neither would be fired until a point in the mission when it was critical that they work. If the descent engine failed, the landing would have to be aborted. If the ascent engine failed, the astronauts on the LM would be trapped on the moon.

Every major component of the Saturn-Apollo "stack" had to be tested, first on unmanned flights and then on manned ones. Testing was vital—the Saturn V alone had more than a million parts, thousands of which could endanger the mission if they failed—but it also ate up time. Von Braun's team had traditionally tested multistage boosters one stage at a time, flying the first stage several times with dummy upper stages before attaching a "live" second stage. Pushed by senior NASA administrator George Mueller, however, they agreed to a new approach called "all-up testing." The first flight of the Saturn V would be made with three "live" stages and a real (though unmanned) Apollo CSM. Mueller, a systems engineer with long experience in ballistic missile design, argued that step-by-step testing took far more time than all-up testing but produced no greater

Figure 7.3: The first orbital test flight of the Apollo LM, part of the Apollo 9 mission of January 1969. Designed to fly only in airless space, the LM had no need for streamlining or equipment for landing on Earth. Here it is flying "upside down," with its docking port at the bottom of the photograph, toward the Earth's surface, and its legs (extended in landing position) at the top. Photograph by Apollo 9 command module pilot David Scott. Courtesy of NASA Johnson Spaceflight Center, image number AS9-21-3212.

confidence in the rocket being tested. The step-by-step method also assumed that any given stage was *likely* to fail. Mueller called on NASA to be optimistic: to expect success and be ready to take advantage of it.

Apollo 4, the first Saturn V test flight, took place on November 9, 1967. It was a brilliant success—the unmanned spacecraft and all three stages of the booster performed flawlessly—and a vindication of Mueller's approach. Apollo 4 also acted as a confidence booster at a time when one was badly needed: six months earlier, on January 27, three astronauts training for the first manned test flight of the CSM had died when fire broke out in the spacecraft during a routine countdown rehearsal. NASA's investigation of the fire revealed more than a thousand flaws in the design of the CSM, hundreds of them serious enough to require immediate correction. The first manned flight of the CSM (aboard a Saturn IB) had been scheduled for February 1967, but the fire, investigation, and redesign postponed it until October 1968. Apollo 4, and the equally successful January 1968 flight of Apollo 5 (an unmanned LM aboard a Saturn IB), helped to bridge the eighteen-month gap. Both flights reassured observers both inside and outside NASA that the program was still moving toward its goal.

Apollo 6, the second all-up test of the Saturn V and the last of the unmanned Apollo missions, was another story. It lifted off flawlessly on April 4, 1968, but things rapidly began to go wrong. A little over two minutes into the flight, the Saturn's first stage began to "pogo": to oscillate violently back and forth along its length, like the bouncing of a pogo stick. The pogoing was brutally fast: five or six cycles every second. Each cycle slammed the rocket forward and then almost instantly backward, applying ten times the force of gravity in each direction. The pogoing continued for ten seconds, then stopped as abruptly as it had started. The first stage finished its burn, and the second stage ignited, without incident. Two minutes into the second stage burn, however, two of its engines quit in rapid succession. Flight controllers corrected as well as they could for the off-balance thrust produced of the remaining engines, but Apollo 6 headed for orbit on a less than perfect trajectory. The third stage ignited on schedule, but flew erratically as its guidance system tried (and failed) to correct the trajectory. It shut down according to schedule, but failed to restart when flight controllers tried to simulate the burn that would send later, manned Apollo spacecraft on their way to the moon. The Apollo 6 CSM reached orbit, but the mission was a partial success at best.

GOING TO THE MOON

The Apollo 7 mission, the first manned test of the CSM since the fire, took place in Earth's orbit in October 1968. It was a complete success, but it used the thoroughly reliable Saturn IB as a booster. The reliability of the Saturn V was still hard to assess. It had performed brilliantly on its first flight (Apollo 4), but suffered from three major breakdowns on its second. Any of the three, if it had happened during a manned mission, would have been grounds for an abort. One (the pogoing first stage) could have injured a human crew if one had been onboard. Engineers had isolated and, they believed, fixed the causes of all three problems, but it would take another Saturn V launch to show if they were right.

The question, for NASA manned space flight chief George Mueller, was what to do with that launch. Mueller's all-up testing philosophy called for each successful test to be followed with a different, more ambitious one. The success of Apollo 7 clearly justified moving on. Apollo 6 might also justify it, but only if the problems that had compromised it had been solved. George Low, the NASA administrator directly responsible for the Apollo spacecraft, suggested sending Apollo 8 to orbit the moon and return to Earth. Other senior NASA officials agreed. A flight to lunar orbit, they argued, would allow astronauts to test navigation, communication, and maneuvering procedures that would be essential for a landing. It would also capture the public's attention in a way that another Earth–orbit mission (something NASA had been doing since February 1962) never could. Mueller saw the value of such a mission, but worried about the risk involved. A spacecraft in Earth's orbit was only a few hours away from a safe landing. *Apollo 8*,[2] when it reached the moon, would be nearly a three days' journey from Earth. An in-flight emergency (like the oxygen tank explosion that crippled *Apollo 13* in 1970) could easily kill all three astronauts.

Mueller gambled, approving the mission, and won. *Apollo 8* lifted off on December 21, spent twenty hours orbiting the moon on December 24–25, and landed safely in the Pacific on December 27. The Saturn booster and the Apollo spacecraft both performed flawlessly, their earlier problems now behind them. The crew brought back invaluable flight experience, high-resolution photographs of potential landing sites, and the first images of Earth as a small, bright blue-and-white disk amid the deep black of space.

The Apollo 9 and Apollo 10 missions served as the final all-up tests of Saturn-Apollo technology. Apollo 9, flown entirely in Earth's orbit in March 1969, tested the handling and life-support systems of a manned LM for the first time. It also demonstrated that the CSM and LM could carry out the rendezvous and docking maneuvers that a lunar landing mission would require. Apollo 10, flown in May 1969, was a nearly complete "dress rehearsal" for the landing attempt. The LM descended to within 10 miles of the lunar surface before firing its ascent engine and returning to rendezvous with the CSM in lunar orbit. When *Apollo 11* lifted off in July 1969, those last 10 miles and the landing itself remained the only untested part of the flight plan. No trainer could simulate, on Earth, what it might

2. All NASA manned spacecraft from *Gemini 4* through *Apollo 8* were "named" for their mission. The name appears in *italics* when referring to the spacecraft, but not when referring to the mission.

be like to land on the moon. The only way to know for sure was to try it. On July 20, Apollo 11 mission commander Neil Armstrong and lunar module pilot Edwin "Buzz" Aldrin did just that.

The success of Apollo 11—a safe landing, Neil Armstrong's "one small step" onto the lunar surface, the placing of scientific experiments, the raising of the American flag, and an uneventful return to Earth—fulfilled President Kennedy's 1961 challenge to the nation. The challenge had been born out of Cold War rivalry, but by 1969 the Cold War had begun to thaw. Richard Nixon, the newly elected president who welcomed the astronauts home, preferred cooperation to confrontation in U.S.-Soviet relations. Soviet premier Leonid Brezhnev sent congratulations rather than boasts. NASA took the same noncompetitive tone in a plaque that Armstrong and Aldrin left at their landing site. The last lines of the inscription read: "We came in peace, for all mankind."

Soviet leaders claimed, after the Apollo 11 landing, that they had never planned to send men to the moon. They insisted that there had been no "race" between the superpowers, and that the whole idea had been an American delusion. The slackening of Cold War tensions in the late 1960s and 1970s made the claim seem plausible to many Americans, but it was a lie. There *had* been a race, and its outcome had been in doubt until the final stages.

THE COMPETITION

Soviet plans for sending men to the moon evolved at the same time, and in much the same way, as American plans. The idea began to appear in Soviet lists of long-range goals as early as 1957–1958, designs for powerful new boosters went on drawing boards in 1960–1961, and a new spacecraft capable of flying to the moon was ordered in 1961–1962 for delivery in 1966. Soviet designers, like their American counterparts, debated the merits of various flight plans before settling on lunar-orbit rendezvous. The Soviets, like the Americans, planned a series of increasingly ambitious flights. Manned test flights of spacecraft in Earth's orbit would pave the way for later flights—first unmanned, then manned—around the moon. The first manned landing would take place in late 1968 or early 1969.

The Soviet lunar program, like earlier Soviet efforts in space, was not run by a single government agency. Instead, government officials chose from among hardware designed and built by separate government-funded "design bureaus." Few of the bureaus were dedicated to rocket design, and few of those with experience in rocket design (through military work) also had experience building spacecraft. This decentralized system led to false starts and duplicate programs that wasted time, money, and resources. The Americans settled on the basic design of their lunar booster and spacecraft by July 1962. The Soviets, on the other hand, continued to design, consider, and discard radically different lunar-landing proposals well into 1965.

The Soviet system also encouraged factionalism and political interference. Sergei Korolev and Vladimir Chelomei, the USSR's two leading rocket designers, headed rival design bureaus throughout the early 1960s. Chelomei had never built rockets for nonmilitary applications, but he guaranteed government interest in his designs by hiring Sergei Krushchev, son of the Soviet premier. Chelomei also formed an alliance with fellow engineer Valentin Glushko, the USSR's premier designer of rocket engines. Korolev's relationship with Glushko had been troubled for decades: Glushko's testimony had helped to send Korolev to a labor camp in the 1930s, and Korolev, Glushko's deputy after World War II, had been promoted past him to become the Soviet space program's "chief designer" in the late 1950s. The new alliance gave Glushko a measure of revenge. Chelomei adopted a powerful new Glushko-designed engine as the basis of his new space booster, the UR-500K. Korolev, meanwhile, had to build his N1 booster around engines designed by Nikolai Kuznetsov, an aircraft-engine designer with no experience in rocketry. The engines worked, but developed little power. Unable to exploit exotic fuels as Glushko did, Kuznetsov traded power for simplicity. In the end, Korolev's N1 went to the pad with thirty engines in the first stage alone.

Chelomei fell from government favor when Khrushchev fell from power in 1964. Official support gradually consolidated, in 1965, behind a modified version of Korolev's plan. Flights to lunar orbit would be made by a stripped-down version of the new Soyuz—a spacecraft designed primarily for work in Earth's orbit—that would carry a single cosmonaut. The landing itself would involve a two-part spacecraft called the L3, launched by Korolev's N1 booster. The larger section of the L3 would remain in orbit with one man aboard while the smaller section carried the other man to the surface. After a postlanding rendezvous in orbit, the cosmonauts would jettison the lander and return home aboard the orbiter. The adoption of these plans was a crucial, momentum-building step for the Soviet lunar program, but NASA had taken the same step three years earlier.

The following year, 1966, should have brought the program's first successes. Instead, it brought disaster. Korolev, admitted to the hospital for what should have been routine surgery, died on the operating table. His successor, Vasili Mishin, lacked his leadership skills and political instincts. Though a competent engineer, Mishin floundered amid the political and administrative crosscurrents that afflicted the Soviet space program. Chelomei, meanwhile, continued to refine his own bureau's lunar-landing plans and propose that they be adopted instead of Korolev's N1-L3 combination. The intrigues failed but damaged the program anyway, squandering its hardwon momentum and eroding Mishin's fragile authority.

Tests of the Soyuz spacecraft, the USSR's equivalent of Apollo, began in late 1966. Almost immediately, new technological problems compounded old organizational ones. Three unmanned test flights failed, one after the other, between November 1966 and February 1967. *Soyuz 1*, the first to fly with a human pilot, crashed on reentry in April 1967, killing veteran cosmonaut Vladimir Komarov. The accident shook the Soviet lunar program just as the *Apollo 1* fire, earlier in the year, had shaken the American program, and no manned Soviet spacecraft flew for a year and a half. Unmanned tests of the L1, a modified Soyuz capable of carrying humans to the moon and back, were little more successful. All four launched in 1967—March, April, September, and November—failed because of problems with their UR-500 boosters. Soviet leaders took stock at the end of 1967. Beating the Americans to the lunar surface now seemed unlikely, but beating them to the first flight *around* the moon was still possible. Focusing on that goal, they pushed the L1 program forward.

Pushing, unfortunately, hurt more than it helped. Overworked engineers and technicians, under pressure to speed up the launch schedule, made errors that destroyed spacecraft or compromised missions. None of the first three L1 test flights in 1968 was a complete success, and only one even reached orbit. These failures, and the Soviets' insistence on two flawless unmanned test flights before the first manned flight, caused the schedule to slip even further. *Zond 5*,[3] the first L1 to leave Earth orbit, did so in September 1968. It flew around the moon and returned a cargo of turtles and insects to a safe landing on Earth. *Zond 6* repeated the flight in November, but its "crew" of animals died when a gasket failure emptied the cabin of air shortly before reentry. Cosmonauts assigned to the L1 program urged Soviet political leaders to let *Zond 7* fly around the moon and back with a human crew in mid-December. Established procedures won out, however, and the mission never materialized. Before the year was over, *Apollo 8* did what *Zond 7* might have done, and spent a day in lunar orbit besides.

The success of *Apollo 8* made it clear to the Soviets as well as the Americans that NASA would attempt a lunar landing in mid-1969. The N1 booster was now ready, however, and a Soviet landing in 1970 or 1971 was

---

3. The name change was deliberately misleading. The earlier "Zond" spacecraft had been small robot probes designed for flyby missions to Venus, Mars, and the moon.

still possible. Hoping, perhaps, for a catastrophic failure that would delay or abort the Apollo program, the Soviets began testing the N1. The results were disastrous: Soyuz and L1 all over again. The first N1, launched in February 1969, was blown up by the range safety officer when a fire broke out in its first stage a minute after liftoff. The second N1 flew for 9 seconds in early July, then fell back onto the launch complex. The explosion that followed destroyed one N1 launch pad and badly damaged the other—damage that would take two years to repair. Two weeks later, Neil Armstrong stepped onto the lunar surface.

The rebuilt N1 launch complex resumed operations in June 1971, but problems with the N1 itself persisted. The third N1 to fly broke up in midair, and the fourth, launched in November 1972, was destroyed by the range safety officer after suffering from fires and exploding engines in the trouble-prone first stage. Soviet hopes of putting cosmonauts on the moon died with the N1. The dismantling of the failed Soviet lunar program began in early 1973. So, ironically, did the dismantling of the successful American program.

APOLLO AS A DEAD END

NASA originally planned nine more Apollo missions to follow the first landing. Apollos 12 to 20 would pursue steadily more ambitious scientific goals, just as Apollos 7 to 11 had pursued steadily more ambitious engineering goals. Their astronauts would stay on the moon longer, spend more time outside the LM, travel farther, do more experiments, and bring back more samples. The later missions would land in more exotic parts of the moon, perhaps even inside a major crater like Tycho or Copernicus, or on the barely explored far side of the moon. Later flights would also carry NASA's first scientist-astronauts: professional scientists who had been trained to fly spacecraft, rather than professional pilots who had been trained in science. Meanwhile, a parallel Apollo Applications Program would use Saturn boosters and Apollo spacecraft to establish, in orbit around the Earth, a modular space station that might one day be expanded to house as many as 100 astronauts.

Had these plans been carried out in full, they might have laid the foundation for a permanent American presence in space. To the intense disappointment of nearly everyone who worked on the Apollo program, they were not. The plans began to erode almost as soon as *Apollo 11* returned safely to Earth. Apollo 20 was cancelled in the first week of 1970. By the end of the year, Apollo 18 and 19 were also gone. Apollo 17 made the final lunar landing in December 1972, carrying geologist Harrison Schmitt, the only scientist–astronaut to walk on the moon. The ever-expanding modular space station at the heart of the Apollo Applications Program became a one-piece, three-man space station called *Skylab*. Lofted into Earth's orbit by the last Saturn V ever launched, it was used by three crews in 1973 and 1974 and then abandoned. Its orbit gradually decayed until, in 1979, it fell into the atmosphere and disintegrated.

One reason for the erosion of NASA's future plans was political. John F. Kennedy, who had initiated the lunar program, and Lyndon Johnson, who had supported it, were gone from the political scene by 1969. Richard Nixon had no political interest and little personal interest in the space program as a whole. Kennedy and Johnson had chosen to spend time and effort promoting Apollo, but Nixon's priorities lay elsewhere. The ongoing war in Vietnam demanded, by 1969, more federal government money than ever before. So, for different reasons, did domestic programs designed to alleviate problems such as poverty, inadequate medical care, environmental degradation, and substandard housing. The annual cost of Project Apollo was modest, even trivial, compared to the annual cost of such programs, but Apollo was easy to caricature and easy to dismiss as a waste of the taxpayers' money.

A second reason for the erosion, often emphasized by veterans of Project Apollo, was public apathy. The later Apollo landings were far more ambitious than the early ones, but the differences were hard for casual observers to see or be interested in. Apollos 7, 8, 10, and 11 had each been a clear leap beyond the mission before. The later Apollo landings seemed—to most of the general public—to be little more than replays of Apollo 11. The later missions drew the public's attention only when something *visibly* different happened: a near-catastrophic explosion on Apollo 13, the first use of the battery-powered "lunar rover" on Apollo 15, or the first nighttime launch of a Saturn V on Apollo 17.

The final, and perhaps most significant, reason was the sense of closure that Apollo 11 brought. President Kennedy had called on the nation to commit itself to "landing a man on the moon and returning him safely to the Earth" by the end of the decade. Apollo 11 had done that, and many Americans saw little reason to continue doing it. The Soviets' abandonment of their own lunar-landing program and the steady improvement of U.S.-Soviet relations reinforced the sense of closure. The "space race" was over, and the United States had won decisively. What, then, was the point of continuing?

The gradual erosion of NASA's once-ambitious plans turned Project Apollo into an operational dead end. It was clear, by the end of 1970, that the commander of Apollo 17 would be the last American on the moon for years to come. There would be no four- or six-man landers, no extended-stay missions, and no permanent scientific stations like those established in Antarctica in the late 1950s. It was also becoming clear that Apollo was a technological dead end. The future of boosters was neither Saturn nor Nova, and the future of spacecraft was not bigger and better versions of Apollo. The future would be a winged, reusable "space truck" that would ride to orbit atop reusable boosters. Officially, it would be called the "Space Transportation System." Unofficially, from the time President Nixon announced it in January 1972, it would be known as the "space shuttle."

Tactical Missiles in the Cold War, 1950–1990

◆

Unguided rockets, designed for short flights along line-of-sight trajectories, have been used in battle for nearly 1,000 years and continue to be used today. They remain potent weapons, capable of doing significant damage if placed in experienced hands and aimed at well-chosen targets. The firefight chronicled in the book (and later movie) *Blackhawk Down* began when a rocket-propelled grenade fired by a Somali militiaman crippled a U.S. Army helicopter and forced it to crash-land in the streets of Mogadishu. The U.S. Army (like most other Western armies) has issued lightweight rocket launchers to its infantry units since the early 1960s. The launchers— a single rocket-propelled projectile stored inside a disposable firing tube— are designed to be fired from the shoulder and discarded after a single use. They are the descendants of the World War II bazooka and *panzershreck*: weapons that, in principle, give a single soldier the power to destroy an enemy tank. No modern tactician, however, would see the foot soldier's unguided rocket as the ideal weapon to use against a helicopter, tank, or strongpoint. For those missions, and a wide variety of others, the guided missile is now the preferred weapon.

The idea of a projectile that could be steered toward its target first received serious consideration during World War I. The first prototypes were built in the 1930s and the first operational models used (mostly by Germany, but also by the United States) in World War II. The development of small guided missiles for the battlefield began in earnest, however, during the 1950s. Advances in solid propellants led to small, powerful rocket motors that could be stored for long periods and required relatively little maintenance, while miniaturized electronic components made it possible to design guidance systems that were both sophisticated and compact. Other changes in military technology also helped to make guided missiles attractive. Aircraft flew higher and faster, tanks were more heavily armored, and warships were better defended than they had been in World War II. Destroying them at close range, with gunfire and bombs, promised to be costly in time, ammunition, and lives. The promised accuracy and destructive power of guided missiles seemed to make them a cheap, efficient alternative. Fired from a safe distance, they would fly unerringly to their target and deliver their lethal payload. Their most optimistic promoters saw them as a "one shot, one kill" weapon that would allow a small force to neutralize (or even rout) a larger and more powerful one.

Soviet and American thinking about tactical missiles focused, in the 1950s and early 1960s, on their uses in a conflict between the superpowers. Antiaircraft missiles, for example, were designed primarily to kill intercontinental bombers: targets that flew high and fast, but on relatively straight courses. Guided missile technology quickly spread, however, to other nations and other military settings. World War III—the full-scale, head-on clash that both superpowers feared but constantly planned for—never happened, but from the mid-1960s to the end of the Cold War, tactical missiles played key roles in a variety of smaller wars. These small wars showed, sometimes spectacularly, what tactical missiles could do. They also showed, sometimes alarmingly, where they fell short of their promoters' dreams.

TACTICAL MISSILE TECHNOLOGY

Tactical missiles vary widely in size, weight, proportions, launch platform, and method of guidance. They are manufactured in a dozen different countries, and designed for a wide variety of missions. Even so, they have many features in common. Nearly all, for example, have the same structure: a metal tube, fitted with stubby fins for steering and stability, which contains a rocket motor, a guidance system, and a warhead. The motor nearly always uses solid fuel, for reliability and ease of storage. The warhead, likewise, is nearly always conventional rather than nuclear. A typical missile's fins are partly or wholly movable. Like the control surfaces on an airplane (rudder, elevators, and ailerons), they alter the machine's course by altering the flow of air around it. The control surfaces in a modern, high-performance airplane move in response to electrical signals carried by wires from the cockpit, where they are generated by the pilot's movement of the controls. The control surfaces in a tactical missile move in response to electrical signals from the guidance system. Missiles, in other words, were using "fly-by-wire" control systems long before such systems became standard in airplanes.

The basic function of a tactical missile's guidance system is to translate input about the relative positions of the (moving) missile and its (probably moving) target into output that will move the missile's control surfaces just enough to keep it on target. The source of the guidance system's input varies widely from missile to missile. Wire-guided missiles, for example, remain connected to their launchers by thin wires as fine as a human hair that unreel behind them and carry inputs from the operator. Radio-guided missiles substituted a wireless link for a wire one, but still required the operator to maintain control. Heat-seeking missiles, on the other hand, use onboard infrared sensors to lock onto and follow the heat generated by the target. Laser- and radar-guided missiles locate and follow their targets using energy (light beams and radio waves, respectively) bounced off the target's surface. Each type of guidance system has both benefits and disadvantages. Radar-guided missiles, for example, can be used in all weather conditions and fired from beyond visual range, but are complex and vulnerable to electronic jamming that disrupts their radar. Heat-seeking missiles are less complex and more reliable, but they can only lock onto their targets at relatively close ranges, and can be lured away from the target by flares and other intense heat sources. Wire-guided missiles are virtually immune to such "spoofing" but oblige the user to remain within visual range of the target until they hit.

Improvements in guidance-system technology gradually improved missiles' performance as the Cold War went on. Early heat-seeking missiles could only "see" and lock onto intense heat sources like jet exhaust pipes and so could only be fired from behind a target aircraft. Later versions, with more sophisticated infrared sensors, could lock onto the friction-heated air streaming back from a target aircraft's nose, and so could be fired from any angle. Early radar-guided missiles depended on an external radar set (part of the aircraft or ground installation firing the missile) to "paint" the target with its signals—often a difficult requirement to fulfill in the chaos of battle. Later versions carried their own radar sets, which allowed them to be used (as heat-seekers had always been) as "fire and forget" weapons. A further improvement supplemented the missile's onboard radar with an inertial guidance system like that used on ballistic missiles. The dual guidance system allowed the missile's radar to be used in brief bursts at critical moments, making enemy jamming more difficult.

Other missile components also evolved, though less extensively, during the Cold War era. "Hot launching," in which the missile's own exhaust sets it in motion, was gradually supplemented by "cold launching," in which a charge of compressed gas (or some other nondestructive mechanism) throws the missile from its launcher before the engine starts. Cold-launch systems allowed missiles to be fired from the decks of ships, the tops of armored vehicles, and the shoulders of infantrymen in confined spaces—anywhere the hot gasses of missile exhaust would cause damage or injury. Missile warheads, originally simple charges of high explosive, also evolved into specialized forms designed for specific types of targets. Armor-piercing warheads carried specially shaped explosive charges designed to focus the force of their explosion inward. Antiaircraft warheads were designed to spray clouds of metal shrapnel through the fragile engines and fuel tanks of their targets. The United States even developed an antiaircraft missile—the Genie—with a nuclear warhead. Designed for use against Soviet bomber formations in the event of an attack on the United States, it was intended to destroy or damage many aircraft in a single blow.

The technological diversity of tactical missiles is much greater than this brief discussion suggests. The range of roles in which they have been used is also broad. Tactical missiles are, by convention, divided into four categories according to where they are launched and where they hit: air-to-air, air-to-surface, surface-to-air, and surface-to-surface. In tracing tactical missiles' impact on warfare, however, it is convenient to blur those distinctions and focus instead on the principal types of targets at which they were aimed.

MISSILES AGAINST AIRPLANES

Airplanes are unique among battlefield targets: hard to hit, but easy to kill. Even the least capable of them can reach speeds measured in hundreds of miles per hour and altitudes measured in tens of thousands of feet. Even the clumsiest can maneuver in three dimensions. Even the most robust airplane, however, is basically fragile: slender structural members and a thin skin wrapped around delicate electronic, hydraulic, and mechanical systems; highly volatile fuel; and a pilot whose only personal "armor" is a cloth flight suit and plastic helmet. Unlike a building, ship, or land vehicle, an airplane cannot be "hardened" against enemy fire except in limited ways and at a terrible cost in performance. Its best defense is its ability to avoid being hit; when that defense fails and a projectile hits, damage is virtually certain. Guided missiles made attractive antiaircraft weapons for two reasons: they increased the chances of a hit, and increased the chances of a hit causing lethal damage.

Both superpowers developed antiaircraft missiles during the 1950s, and both did so with the problem of homeland defense firmly in mind. ICBMs were still in their technological infancy in the late 1950s—crude, inaccurate, and unreliable—and the piloted bomber was still the weapon of choice for long-range strategic attacks. The defensive plans of both superpowers were built, therefore, around the problem of stopping long-range bombers, and antiaircraft missiles were designed primarily with bombers in mind. Early surface-to-air missiles (like the Soviet-made SA-2) and air-to-air missiles (like the American-made Sidewinder) tended to be optimized for a specific mission: destroying big airplanes flying relatively straight-and-level courses at relatively high altitudes. Designers also assumed that the missiles would be stored and maintained at well-equipped, permanent airbases, and deployed only in the face of an imminent threat. These assumptions made the problem of designing a successful missile seem less difficult than it might have, and encouraged optimism on the part of both aerospace contractors and military customers. Firing tests, often designed to maximize the missile's chances of success, reinforced the optimism.

Two early uses of antiaircraft missiles in the "real world" heightened expectations even further. On September 24, 1958, pilots from the Communist and Nationalist Chinese air forces met in the skies over the disputed islands of Quemoy and Matsu. The Communist pilots had numbers and experience on their side, but the Nationalist pilots had technology: brandnew, heat-seeking Sidewinder missiles mounted beneath the wings of their F-86 fighters. Nationalist pilots destroyed ten enemy aircraft in a matter of minutes, while losing none of their own. The Communists, startled by a weapon they could neither match nor evade, broke off the battle and ceded the airspace over Quemoy and Matsu to the Nationalists. Less than two years later, a single SA-2 surface-to-air missile knocked an American U-2 reconnaissance plane out of the sky over the Soviet city of Sverdlovsk. The plane crashed, the pilot ejected, and the Soviets triumphantly displayed both to a world that had just heard President Dwight Eisenhower categorically deny their existence. The loss of the plane and the capture of CIA-trained pilot Francis Gary Powers was a diplomatic black eye for the United States, but it was a revelation for military leaders interested in the problem of air defense. The U-2 was designed to fly so high that no antiaircraft weapon could touch it, but a surface-to-air missile had brought it down with ease (another SA-2 would destroy another U-2 in October 1962). If the new missiles could touch even the "untouchable" U-2, they could easily deal with the bigger, lower flying bombers.

Fueled by this optimism, both superpowers developed a wide range of antiaircraft missiles in the late 1950s and early 1960s: air-to-air as well as surface-to-air, radar-guided as well as heat-seeking. Just as important, they increased their reliance on guided missiles in air defense and aerial combat. The Soviet Union pioneered a new style of air defense that integrated guns, surface-to-air missiles, and radar installations into a centrally controlled network. Over the next two decades, they would export both the concept and the technology to their allies and client states: North Vietnam, Egypt, Syria, and Iraq among them. Both superpowers made air-to-air part of the standard missiles of their fighters' armament, and the United States went further still: introducing fighters armed solely with missiles. The first missile-only fighter, the Air Force's F-102 Delta Dagger, entered service in 1956. The Navy's F-4 Phantom, later adopted by the Air Force as well, followed in 1961. Most American fighters of the period carried both missiles and cannon, but planners and pilots alike expected missiles to be their primary weapon. Why close to 2,000 feet or less (the maximum effective range for cannon) if you could fire from 3,000 feet (the minimum effective range for missiles) or more? The American preference for missile-armed airplanes reflected the American military's faith in missiles' effectiveness. Studies in the early 1960s confidently predicted that, in combat, the radar-guided Sparrow would hit its target 65 percent of the time and the heat-seeking Sidewinder would hit 71 percent of the time.

The Vietnam War put both superpowers' missiles—and assumptions about them—to their first real test. North Vietnam's air defenses in mid-1964 consisted of three dozen jet fighters, 1,000 antiaircraft guns, a few radar stations, and no missiles at all. After the Gulf of Tonkin incident in August 1964, however, the Soviet Union began to supply missiles, launchers, and additional radar sets, as well as advisors, to oversee their construction and train their North Vietnamese crews. By the time Operation Rolling Thunder—U.S. air attacks on North Vietnam—began in March 1965, North Vietnam had a Soviet-style integrated network of missiles, guns, and radar with which to meet them. The United States, meanwhile, deployed large numbers of missile-armed fighters in combat for the first time: Navy F-4 Phantoms and F-8 Crusaders from aircraft carriers in the South China Sea, and Air Force F-4s from bases in Thailand. Soviet-built MiG-21s, deployed in early 1966, gave the North Vietnamese Air Force their own missile-armed fighters: each carried two AA-2 heat-seeking missiles (similar to the Sidewinder) in addition to cannon.

The air war over Vietnam lasted, with only a few brief pauses, from 1965 until 1973. It was the first full-scale war in which both sides had used missiles against aircraft, and the first sustained use of missiles in combat. On one hand, it confirmed what the Chinese dogfight of 1958 and the U–2 shoot-downs of 1960 and 1962 had suggested: missiles could be devastatingly effective against undefended airplanes flown by unaware pilots. On the other hand, however, it showed that neither condition was likely to persist in wartime. Both sides were at first caught unprepared by the missile threat, but both sides (the United States and Soviet-backed North Vietnam) quickly moved to prepare better defenses. Both sides, too, quickly learned the limitations of the other's missiles. Tactics changed, new countermeasures entered service, and the combat effectiveness of missiles diminished.

American air-to-air missiles, in particular, were dogged by performance problems. Designed for use against high-flying, slow-moving bombers, they proved to be badly adapted for use against low-flying, highly maneuverable fighters. Both the Sidewinder's infrared sensors and the Sparrow's radar guidance system were prone to lose track of low-flying targets amid the "noise" of other heat sources and radar echoes from the ground. Both missiles were complicated to fire, neither could be fired within 3,000 feet of a target, and both had trouble following enemy fighters through tight turns. North Vietnamese pilots adapted their tactics to these shortcomings. When they spotted a missile (especially easy with the Sparrow, because of its prominent smoke trail), they turned sharply and dove for the treetops. Both missiles—the Sparrow in particular—also suffered from reliability problems. Tropical humidity and salt air damaged electronic circuitry, rough roads and carrier-deck landings jarred fragile systems into failure, and maintenance took place under less-than-ideal conditions with limited tools and supplies.

Prewar forecasts predicted two hits for every three air-to-air missiles fired. Wartime results, however, fell well short of that goal. Air Force pilots attempted to fire Sparrows sixteen times in April and May 1965; three failed to fire, and only one of the remaining thirteen hit its target—a 6 percent hit rate that improved only marginally by 1968. Navy crews fired twelve Sparrows in the same two-month period, achieving four hits for a 25 percent hit rate—a significant improvement on the Air Force's record, but still well short of expectations. Six of twenty-one Sidewinders fired during April and May hit their targets (28 percent). One squadron commander, responding to his pilots' frustration with missile performance in 1965, said: "Guys, they don't call them hittles" (Michel 1997, 44). Incremental upgrades were made to both Sparrow and Sidewinder during the war, but they did little to improve performance. During the last years of direct U.S. involvement in Vietnam (1971–1973), the hit rate for both Sparrows and Sidewinders hovered just below 12 percent and that for the Falcon—a new heat-seeking missile ordered by the Air Force as a successor to the Sidewinder—around 16 percent (two hits in twelve launches). The Navy's new "G" model of the

Sidewinder was the only bright spot: its twenty-three hits in fifty firings gave it an astounding (for the time) hit rate of 46 percent.

The bomber crews and mission planners responsible for attacks on North Vietnam were, simultaneously, learning the weaknesses of North Vietnamese surface-to-air missiles (SAMs). The SA-2 was big—often compared to a flying telephone pole—and relatively unmaneuverable. Like the Sparrow and Sidewinder, it had been designed to destroy big, sluggish bombers at high altitude. American pilots adjusted their tactics to exploit these weaknesses, flying low on their runs to and from the target, and "popping up" at the last moment to release their bombs. Mission planners also made adjustments to deal with the SAM threat. Attack aircraft carried radar-jamming equipment along with bombs, and specialized radar-jamming aircraft accompanied raids. Specially equipped aircraft dubbed "Wild Weasels" led the raids, attacking enemy missiles sites with bombs, gunfire, and new types of radiation-seeking missiles that homed in on SAM sites' targeting radar. A typical late-war mission, flown on May 10, 1972, consisted of thirty-two bombers escorted by twenty-eight fighters to defend against MiGs and twenty-seven escorts (jammers and Wild Weasels) to defend against SAMs.

The arrival of Soviet-made SA-7s on the battlefield in 1972 forced American pilots and planners to change their tactics again. A lightweight, heat-seeking missile designed to be carried and fired by a single soldier, the SA-7 was effective where the SA-2 was not: at ranges under 2 miles and altitudes below 10,000 feet. Low-level attacks became far more dangerous and, as a result, far less common. Modern jets like the F-4 and the F-105 Thunderchief took to higher altitudes, trusting jamming equipment, Wild Weasel escorts, and their own maneuverability to keep them safe. Slower, less maneuverable aircraft like the aging, propeller-driven A-1 Skyraider were pulled off the battlefield. Helicopters, though extremely vulnerable to SA-7s (it took an average of 135 SA-7s to kill an F-4, and 10 even for an A-1, but only 1.8 for a helicopter) remained in service. They were too valuable to the U.S. war effort to consider withdrawing them.

The Vietnam War demonstrated both the power and the limitations of antiaircraft missiles. The Arab-Israeli war of 1973 reinforced those lessons. Egypt and Syria, whose air forces had been destroyed on the ground by Israeli air strikes at the start of the 1967 war, fortified strategic areas with belts of antiaircraft defenses—integrated networks of radar, missiles, and guns like those used by North Vietnam. Egypt, for example, deployed 150 SA-2 and SA-3 batteries, dozens of mobile SA-6 batteries, and thousands of guns and shoulder-fired SA-7s. Sixty batteries were allotted to the Suez Canal alone. The Israeli Air Force lost 115 aircraft (one-third of its prewar strength) in nineteen days of fighting, all but four to the Arab nations' ground-based air defense networks.

Eight years of relative peace (1973–1982) followed the American with-drawal from Vietnam and the negotiated truce between Israel and its neigh-bors. Over those eight years, a new generation of missiles reached service. Deployed in a series of conflicts in the early 1980s, they showed that both antiaircraft missiles had reached a new level of sophistication.

On August 19, 1981, a pair of Libyan combat aircraft threatened ele-ments of the U.S. Navy's Sixth Fleet operating in international waters off the coast of Libya. A pair of F-14 Tomcat fighters from the aircraft carrier *Nimitz* quickly destroyed both Libyan intruders, using upgraded versions of the once-troubled Sparrow. An upgraded version of the Sidewinder served the British well in their war with Argentina the following year: of twenty-six missiles fired in twenty-three engagements, roughly 90 percent hit their targets. Estimates of Argentine losses to Sidewinders varied from sixteen to twenty aircraft, but even the lowest of those figures gave the Sidewinder a kill ratio of 61 percent—five times the U.S. average in Vietnam a decade earlier. Israel's 1982 attacks on guerilla bases inside Syria produced similar results: eighty-four Soviet-built Syrian fighters destroyed with Sparrow and Sidewinder missiles, including sixty-four in the first two days of fighting alone. The "one shot, one kill" dreams of 1950s missile designers had be-come reality, and missiles had displaced guns as the dominant weapon in aerial combat.

Antiaircraft missiles fired from the ground also achieved fresh successes in the 1980s, notably during the Soviet occupation of Afghanistan (1979–1989). Agents of the Central Intelligence Agency supplied hundreds of U.S.-made Stinger missiles to Afghan guerillas beginning in mid-1986. The Stinger, a shoulder-fired, heat-seeking weapon with a range of 3–5 miles, inflicted devastating losses on low-flying Soviet aircraft—especially the heavily armed helicopter gunships that had been the Soviets' most potent weapon. Soviet commanders pulled the helicopters back to higher altitudes in response, reducing their vulnerability but also their effectiveness. The po-tent, portable Stinger fitted perfectly with the guerillas' hit-and-run strat-egy, and the losses it inflicted reinforced the Soviet government's decision to withdraw.

MISSILES AGAINST SHIPS

The crude antiship missiles used in World War II evolved steadily in subse-quent decades, acquiring more efficient motors, more accurate systems, and more powerful warheads. Warship designers countered with improved defensive measures, but few of these systems—offensive or defensive—were ever tested in combat. The armed conflicts of the Cold War took place almost exclusively on (and over) land, and rarely involved *two* powers capable of challenging one another for control of the sea. The 1982 war between Britain and Argentina over the Falkland Islands was an exception. British forces, operating thousands of miles from home in the remote South Atlantic, depended on ships for transportation, supply, and (most critically) air support. Without their fleet of transports, aircraft carriers, and escorts, the British could not hope to retake the islands from occupying Argentine forces. Argentine air force commanders, well aware of this fact, placed British ships high on their list of desirable targets. British naval commanders, equally well aware of it, deployed their forces with extreme caution. They refused, for example, to send their carrier battle groups into the waters between the Falklands and the Argentine mainland—reducing their effectiveness, but also reducing the chances that one of the vital carriers would be lost.

Both sides used antiship missiles in the Falklands War, and both sides were stunned by their power. On May 3, 1982, helicopters launched from the decks of British destroyers used Sea Skua missiles to attack a pair of 700-ton Argentine patrol boats. Within minutes, the *Commodoro Somellera* was sinking and the *Alferez Sobral* badly damaged and afire. The leader of the attack, Lt. Commander Alan Rich, returned to his ship "shaking uncontrollably, overwhelmed by the destructive power he had unleashed" (Hastings and Jenkins 1984, 151). A day later, Argentine pilots exacted revenge: a French-made Exocet missile, fired at close range and wave-top height, slammed into the side of the British destroyer *Sheffield*. The missile tore a 10-by-4-foot hole in the hull, set the ship afire, destroyed the water main, and knocked out electrical power. Twenty-one men died as a result of the attack, and the surviving crew, deprived of any means of fighting the fire, were forced to abandon the ship. *Sheffield*, fires finally burnt out, was taken under tow five days after the attack, but sank before it could reach a friendly port. "Everybody had always said that modern warships are 'one-hit ships,'" *Sheffield*'s captain observed after the attack. "Nobody had thought about the implications of a 'one-hit ship' 8,000 miles from home. It's much worse than a car crash—you lose everything" (Hastings and Jenkins 1984, 155).

Two weeks later, on May 25, another Exocet struck the cargo ship *Atlantic Conveyor*—killing twelve sailors and sending it to the bottom along with nearly all its cargo. The cargo lost included thirteen of the fourteen helicopters aboard, and all the tents for the invasion force. It was a major strategic setback, but it could—British commanders soberly reflected—have been even worse. Losing the *Atlantic Conveyor* simply delayed the invasion; losing one of the fleet's two aircraft carriers would have forced a complete withdrawal. The carriers were pulled back even further.

The "tanker war" that raged in the Persian Gulf in 1984–1987 confirmed the destructive power of antiship missiles. Part of a larger, decade-long conflict between Iran and Iraq, it resulted from each country's desire to cut off the other's oil exports and so strangle its economy. The tanker war included 286 recorded missile attacks, most executed by Iraqis armed with French-made Exocets. More than 100 ships suffered significant damage, and eighty were sunk, scuttled, or written off as total losses. The U.S. Navy frigate *Stark* was one prominent victim of the tanker war. Hit by two Iraqi Exocets while escorting tankers through the Gulf in May 1987, it sustained serious fire and blast damage and suffered thirty-seven dead, but was able to reach the friendly port of Bahrain under her own power. The attack on the *Stark*, which had failed to detect the incoming missiles until a split second before impact, heightened tensions aboard U.S. naval vessels in the Gulf. A year later, in July 1988, the cruiser USS *Vincennes* fired two missiles at an unidentified aircraft whose course changes and radio silence suggested that it was preparing another attack. The aircraft—which one of the missiles hit and destroyed—was in fact an Iranian airliner with 290 civilians aboard.

The Falklands War and Iran-Iraq wars left a deep impression on naval officers. The advent of missiles costing hundreds of thousands of dollars that could sink ships costing tens of millions led, beginning in the mid-1980s, to steadily more sophisticated shipboard defenses. Warship designers upgraded chaff dispensers[1] (British ships in the Falklands had carried only seven rounds of chaff apiece), and added automated gun and missile systems to destroy incoming missiles at close range. The *Vincennes* was, ironically, the first of a new class of American "air defense cruisers" designed to defend an entire battle group by engaging up to 200 missiles at once. Designers also began, in the late 1990s, to use "stealth technology"—using new shapes and materials to make ships less visible to radar. How effective these measures will be against a new generation of antiship missiles remains to be seen. Large-scale naval combat has, since 1991, become a rarity again.

---

1. Chaff is strips of metal foil or metal-coated plastic film, which reflect radar signals. Dropped or thrown into the air around a would-be target it can confuse radar-guided missiles by overwhelming their guidance systems with thousands of false "contacts."

MISSILES AGAINST TANKS

The assumption that only a tank could destroy a tank was undermined, in World War II, by the introduction of antitank artillery (like Germany's versatile 88-millimeter cannon), mines, hand-carried "satchel charges" of high explosive, and unguided rockets like the American bazooka. All three types of weapon offered some defense against armored attacks, but each had drawbacks. Artillery had to be sited, and mines sown, in advance; they were effective in defense of fixed points, but less so in dealing with unexpected breakthroughs by enemy tanks. Satchel charges and unguided rockets could be used without extensive preparation, but only at close range and without any guarantee of lethal damage to the target. Antitank missiles, pioneered by French designers in the mid-1950s, combined the portability and flexibility of rocket launchers with the accuracy and hitting power of artillery. Carried by small teams of soldiers or mounted on wheeled vehicles, they allowed Cold War–era infantry to engage enemy tanks on something like equal terms.

Technologically, antitank missiles were less complex than antiaircraft or antiship missiles. Unlike aircraft, tanks are slow-moving targets capable of maneuvering in only two dimensions. Unlike ships, they have no means of detecting incoming missiles and no defensive weapons to use against them. Missile designers assumed that most antitank missiles would be fired at relatively close range by operators who could follow their progress and could adjust their course as needed. Effective guidance for antitank missiles was, therefore, a relatively straightforward problem: one solved initially with optical guidance systems and later with infrared, laser, and wire-guidance systems. The French SS-11, introduced in the late 1950s, automated the process of adjusting the missile's course. The operator kept the launcher's sights centered on the target, and the guidance system adjusted the missile's course as needed. Warhead design was also straightforward, since missile designers could easily adapt armor-piercing techniques developed for artillery shells.

Antitank missiles became a dependable, effective weapon more quickly than antiaircraft or antiship missiles. As a result, they were the first class of tactical missile to have a significant impact in combat. Israeli forces used French SS-10s to destroy Egyptian tanks in the Sinai Desert during the 1956 Arab-Israeli war, and two years later Britain's secretary of state for war announced that a new missile then in development would "remove the heavy tank from the battlefield" (Gunston 1993, 248). The prediction was premature, but the United States, the Soviet Union, and their European allies invested heavily in antitank missiles during the 1960s and early 1970s, preparing for the massive tank battles that were expected to take place on the plains of central Europe if a U.S.-USSR war broke out. Antitank missiles also continued to play roles in smaller wars outside Europe. Both Arab and Israeli forces used them in the Six-Day War of 1967, and North Vietnamese forces used Soviet-made AT-3s against American tanks during their spring 1972 invasion of South Vietnam.

It was the 1973 Arab-Israeli War, however, that established antitank missiles as a major force in armored warfare. The missiles had, by 1973, achieved "one shot, one kill" capability, and both sides used them with devastating effect. Israeli forces armed primarily with American-made TOW (*t*ube-fired, *o*ptically tracked, *w*ire-guided) missiles destroyed 1,270 Egyptian and Syrian tanks—more than a third of the two nations' combined tank forces. In the Sinai Desert, however, Egyptian forces enjoyed even greater success. Israel had deployed 290 tanks to the Sinai, the southern theater of a two-front war; Egyptian troops armed with Soviet AT-3s and rocket-propelled grenades destroyed 180 of them—a staggering 62 percent. Observers on both sides of the war concluded that missiles, not other tanks, were now the most potent threat to armored forces. The world's major armies adjusted their plans accordingly in the late 1970s and 1980s—searching for ways to make their antitank missiles more deadly and their tanks less vulnerable.

CONCLUSION: A NEW BATTLEFIELD

The basic goal of warfare has not changed since the days that Homer wrote about in the *Iliad*: inflict enough damage on the enemy's troops to achieve your goal, while exposing your own troops to the least possible risk. The evolution of military technology reflects those goals. Weapons have grown steadily more powerful and accurate (which makes them more efficient) and steadily more able to strike at a distance (which makes their users less vulnerable). Most advances in military technology are small, incremental changes that have only a modest impact on strategy and tactics. Occasionally, however, a new type of weapon transforms both. The airplane (which added a third dimension to the battlefield) was one such weapon. The tank (which combined the mobility of cavalry and the destructive power of artillery) was another. The tactical guided missile was a third.

Even the earliest tactical missiles offered a powerful combination of attributes: destructive power, portability, and relatively low cost. Their destructive power made aircraft, ships, and tanks—the basic tools of mid-twentieth-century warfare—more vulnerable than ever before. Their portability and low cost allowed them to quickly spread beyond the boundaries of the industrialized nations that built them. The United States and the Soviet Union, arming their allies against one another, shipped huge quantities of their missiles abroad. Missiles also became important in the international arms trade, with France (for example) supplying its Exocet anti-ship missile to the air forces of Argentina, Iraq, and other countries. Tactical missiles had, by the time the Cold War ended in 1990, spread to every nation where political tensions led (or threatened to lead) to military action: Taiwan, Korea, Vietnam, Israel, Lebanon, Iran, Iraq, Afghanistan, and so on. When war did break out in those areas, it was shaped by the widespread presence of the missiles. Nations that used them effectively (Egypt and Syria in 1973, Israel in 1982) enjoyed a great advantage over nations that did not. Even the United States (in Vietnam) and the Soviet Union (in Afghanistan) found themselves put on the defensive by enemies that had mastered the use of tactical missiles.

Tactical missiles transformed the battlefields of the Cold War era because of the threat they posed to expensive, powerful military "assets" like planes, ships, and armored vehicles. The simple fact that the missiles existed—that they *might* be used—forced the commanders who controlled those assets to deploy them more carefully. Ships approached enemy shorelines less frequently, and treated every unidentified aircraft as a potentially lethal threat. Bombers and attack aircraft flew high to avoid the threat of missiles, sacrificing surprise and accuracy in order to avoid crippling losses. Armored vehicles moved more cautiously, wary of opportunistic missile attacks by enemy infantry and aircraft. Commanders advancing boldly into enemy territory risked disaster if the enemy's missiles were not neutralized first—either by preemptive attacks or by defensive measures like flares, chaff, jamming, or evasive maneuvers.

Defensive measures grew steadily more sophisticated and effective as the Cold War went on. So, not surprisingly, did the missiles themselves. Warheads became more powerful, guidance systems more capable, and missiles as systems more reliable. The threat of enemy missiles could thus be reduced, but never eliminated. Effectively using (and defending against) missiles had, by 1990, become as important to military success as controlling the high ground.

9

Spaceflight Becomes Routine, 1970–Present

◆

The Soviet and American space programs of the 1950s and 1960s were government-run operations designed to enhance national prestige and pre-serve national security. They enjoyed, as a result, lavish support from the governments that sponsored them: money, land, personnel, equipment, and publicity. The winding down of the "space race," the thawing of the Cold War, and the rise of pressing domestic problems in both nations caused that support to erode rapidly in the early 1970s. Both superpowers remained committed to their space programs, but neither was willing to continue the headlong rush that *Sputnik* had begun in 1957.

The Soviet and American space programs of the 1970s and 1980s were designed to turn space travel from a spectacle into a routine event. Both were intended to make the presence of humans in Earth orbit common-place. Both, finally, were meant to lower the cost of launching payloads into space. The Soviet program pursued these goals by exploiting proven tech-nologies from the 1960s. The cosmonauts who served aboard the *Salyut* and *Mir* space stations flew to orbit aboard Soyuz spacecraft and were launched by standardized versions of the well-tested Soyuz launch vehicle. The American program, on the other hand, poured its limited resources into an entirely new, tightly integrated combination of spacecraft and booster designed to make space launches truly affordable. The space shuttle was unlike anything that NASA had flown before. It was, in several critical ways, as much of a step beyond Apollo/Saturn as Apollo/Saturn was beyond Mercury/Redstone.

The Soviet and American space agencies were not, however, the only groups struggling to make space travel routine. The European Space Agency (ESA) designed, built, and in 1979 began launching its own Ariane rockets from a site on the Caribbean coast of South America. The success of Ariane radically altered the world of space launches by making a two-player game into a three-player game. All three players—the United States, USSR, and ESA—succeeded, to varying degrees, in making space launches routine. None, however, succeeded in quite the way they had anticipated.

THE SOYUZ-U

On May 18, 1973, a Soviet rocket lifted off from the Plesetsk launch site on a mission to Earth orbit. It was the first of a new type, the Soyuz-U, and over the next thirty years it would become the most-used launch vehicle of the space age. The Soyuz-U did for the Soviet space program what the Ford Model T did for American automobile owners and the Douglas DC-3 did for the world's airlines: set a new standard for durability, versatility, and reliability.

The Soyuz-U reflected Soviet rocket engineers' preference for incremental improvement over radical innovation. It was a refined, improved version of the basic Soyuz launcher introduced in 1966, and thus a direct descendent of the R-7 missile that had launched *Sputnik* in 1957. The Soyuz-U shared the engines and configuration of the basic Soyuz launch vehicle. Four strap-on boosters, each with a single four-chamber engine burning kerosene and liquid oxygen (LOX), surrounded a two-stage "core" (see Figure 9.1). The boosters and the first stage of the core (which used a similar engine) would fire simultaneously at liftoff, and in the next 4.5 minutes provide most of the thrust necessary to reach orbit. The second stage, using a smaller four-chamber kerosene-LOX engine, would then ignite, lifting the payload into orbit. The critical difference between the basic Soyuz and the Soyuz-U was the newer launcher's use of a chilled, high-density fuel in the core stages. The new fuel significantly improved the engines' efficiency, allowing the Soyuz-U to carry larger payloads and reach higher orbits.

The Soyuz-U was intended, like the basic Soyuz it replaced, to be the Soviet space program's standard launch vehicle: a versatile workhorse capable of lifting satellites, space probes, or manned spacecraft. It succeeded brilliantly. Thirty years have passed, at this writing, since the first Soyuz-U launch at Plesetsk. Over those thirty years, 698 more Soyuz-U launchers

Figure 9.1: A Soyuz-U launch vehicle on the launch pad at Baikonur, Kazakhstan, preparing to launch the Soyuz spacecraft that would take part in the Apollo-Soyuz orbital rendezvous of 1975. A descendent of the R-7 missile, the Soyuz-U is the most widely used and most reliable manned launch vehicle in history. Courtesy of NASA Headquarters, image number 75-HC-606.

lifted off the pads at Plesetsk and Baikonur, carrying every type of payload in the Soviet (later Russian) space program's inventory. Only eighteen of those 699 launches have ended in failure, giving the Soyuz-U an astonishing success rate of 97.2 percent.

The cost of developing the Soyuz-U was relatively low, since it was modified from a well-tested design. The cost of operating it was also relatively low, since it could use launch and maintenance facilities built for earlier members of the R-7 family of rockets. Support crews, launch technicians, and engineers familiar with the older Soyuz could apply most of their knowledge and experience directly to the new model, which shared its engines and many other components. The Soyuz-U was thus an extremely cost-efficient way to put payloads in orbit—cheap to build and (as single-use launch vehicles go) cheap to operate. The Soviets could make dozens of launches a year through the 1970s and 1980s because they had an economical, highly reliable launch vehicle at their disposal.

The Soyuz spacecraft that carried Soviet cosmonauts to orbit throughout the 1970s and 1980s reflected the same philosophy that produced the Soyuz launcher. It was (and remains today) essentially the craft that Korolev designed in the mid-1960s for missions to the moon. Like the Soyuz launch vehicle, the Soyuz spacecraft was gradually modified but never wholly redesigned. After a catastrophic air leak killed the crew of *Soyuz 11* in 1971, for example, one of the Soyuz's three seats was removed in order to give the remaining two crew members room to wear pressure suits and helmets. The Soyuz-T (first flown in 1980) and the Soyuz-TM (first flown in 1986) introduced further upgrades to the basic design: computerized flight controls, an escape rocket to pull the spacecraft free of its launch vehicle, and a variety of smaller changes. Reductions in the weight and volume of equipment made it possible, by the TM model, to raise the crew complement to three pressure-suited astronauts. The Soyuz spacecraft, like the Soyuz launch vehicle, proved itself both versatile and reliable. It made dozens of flights in the 1970s and 1980s: ferrying the crews and cargo of the *Salyut*, *Almaz*, and *Mir* space stations as well as flying solo missions, and, in 1975, the Apollo-Soyuz rendezvous that marked the symbolic end of the space race. Like its launch vehicle namesake, it remains in use today.

THE SPACE SHUTTLE

The history of the space shuttle is one of engineering triumph interwoven with political tragedy. The triumph is that, in less than a decade, NASA and the American aerospace industry built and successfully flew the world's first reusable spacecraft. The tragedy is that, in a time of shrinking budgets, NASA won approval for the shuttle program only by making extravagant promises that the shuttle could not fulfill. Judged as what it really is—the experimental prototype for a new generation of spacecraft—the shuttle has been a great success. Judged as what NASA promised Congress it would be—a "space truck" that would make access to Earth's orbit cheap and routine—it has been a dismal failure.

The shuttle was originally conceived, in 1968–1969, as one element in an array of spacecraft and space stations. Stations would orbit both the Earth and the moon, and a nuclear-powered "transfer vehicle" would carry passengers and cargo back and forth between them. A "space tug" powered by chemical rockets would move payloads from low Earth orbit to higher, geosynchronous orbits. A specialized lunar lander would travel between the lunar surface and the station in lunar orbit. The space shuttle was thus part of a larger system: the link between Earth and the Earth-orbiting station, and the deliverer of satellites to the tug.

The shuttle was also, in the original 1968–1969 plans, a system in its own right: a fully reusable orbiter launched into space atop a fully reusable booster. The booster, like the orbiter, would have wings, pilots, and two sets of engines—rockets for the "outbound" leg of the flight, and jets for the return flight through the atmosphere. Carried into the upper atmosphere by the booster, the orbiter would then fire its own engines and proceed to orbit to carry out its mission. The booster, meanwhile, would fly back to the launch site for refueling and refurbishing. By the time the orbiter returned to the launch site, the booster would nearly be ready for another mission.

These ambitious plans suffered, between 1969 and 1972, from the same budget reductions that brought an early end to Project Apollo. The space stations and other spacecraft were eliminated, leaving the shuttle without a destination or a function. The design of the shuttle itself was also pared down, and the fully reusable piloted booster exchanged for a disposable external fuel tank and two solid-fuel strap-on boosters. The decision made the shuttle cheaper to build (one piloted vehicle instead of two), but more expensive to operate (the SRBs would have to be refurbished, and the external tank replaced, after every flight). It also meant that the shuttle was, more than ever, an integrated system. The shuttle's external tank and SRBs were simply an adjunct to the orbiter—not an independent booster that could be used independently for other missions (see Figure 9.2). NASA's concern with cost control was also evident in other design changes made during the 1970s. The jet engines originally planned for the orbiter were dropped, as was the planned crew escape module, and a system for shutting down the

Figure 9.2: The space shuttle *Atlantis* climbs away from Kennedy Space Center on the STS-45 mission of March 1992. The shuttle is powered, at this stage of its flight, by the orbiter's three main engines (arranged in a triangle below the tail fin) and the two large, white solid rocket boosters flanking the fuel tank beneath the orbiter. The two smaller engines of the orbital maneuvering system (OMS) are contained in the bulges flanking the tail fin. Their exhaust nozzles are visible to either side of the uppermost main engine. Courtesy of NASA Kennedy Space Center, image number 92PC-0644.

SRBs in an emergency. All the changes lowered development costs and—by reducing weight and complexity—lowered operating costs as well.

Lowering costs was critical because, with the cancellation of the space station, NASA had promoted the shuttle as an economical, all-purpose launch vehicle capable of putting payloads into Earth orbit for $20–$50 a pound. A skeptical Nixon administration had approved the shuttle on that basis in 1972, reasoning that (if NASA's predictions were correct) the shuttle would make disposable launch vehicles like Atlas, Titan, and Delta obsolete. NASA feared, probably with good reason, that if the development costs of the shuttle soared, the entire program would be a target for cancellation.

The shuttle, in its final form, was an ungainly-looking product of design by committee, shaped by hundreds of compromises both large and small. It was also, however, a spectacular feat of engineering—one that used, but went well beyond, technological elements developed for Apollo. The shuttle's main engines burned liquid hydrogen and liquid oxygen like those of the Saturn-series boosters, but far more efficiently. The orbiter's two orbital maneuvering engines, used to change orbits and decelerate the ship for reentry, used ignition-on-contact fuels similar to, but more powerful than, those burned in the Apollo LMs. Every square foot of the orbiter's curved skin was protected by individually molded silica tiles, each formulated for the heat that its particular area of the skin would experience during reentry (see Figure 9.2). The orbiter's computer was designed to be capable of guiding the ship, without human input, through the long glide from orbit to safe landing.

The solid rocket boosters were a substantial engineering achievement in their own right. There had been large solid-fuel rockets before—Polaris, Minuteman, and Titan III—but the shuttle SRBs would be even larger: 125 feet from nosecone to nozzle, and 12 feet in diameter. They also had to be built so that, after recovery, they could easily be refurbished and refueled for their next flight. Three of the four SRB designs submitted to NASA in 1973 solved these problems in basically the same way: by building the booster in segments that would be assembled at the launch site to form a complete rocket. The joints between these segments—called "field joints" because they would be put together "in the field," away from the factory—consisted of a steel rim (the "tang") on the top edge of the lower segment that fitted into a deep groove (the "clevis") on the bottom edge of the upper segment. To keep hot gasses from escaping through the joints during flight, each joint was fitted with a pair of slender rubber seals (the "o-rings") protected by heat-resistant zinc putty. Like the basic segmented design, the o-ring seals had been used before on the Titan III launch vehicle. Aware that human lives would be riding on the performance of the SRBs, the designers used two o-ring seals in each joint, instead of the single seal used on the Titan III.

Reliability was the single greatest challenge that the SRB designers faced. Unlike liquid-fuel engines, solid rocket motors could not be shut down or throttled back in flight. Given the design, NASA officials originally included an emergency shutdown system in its specifications for the SRBs (explosive charges would blow out the forward end of the casing, venting the hot gasses and killing the rocket's thrust), but concerns about weight and potential damage to the orbiter led them to abandon it. Ignited on the pad at liftoff, the boosters would burn at full power until their fuel was exhausted two minutes later and 150,000 feet higher. The lives of the shuttle crews would depend on them working, flawlessly, every time.

The first orbital test flight of the space shuttle took place in April 1981, a little more than nine years after the project began in earnest, and three years later than originally planned. The Saturn V booster and Apollo spacecraft had, in an era of far more generous NASA budgets, taken just under eight years from their approval in May 1961 to their first full test flight (Apollo 9) in January 1969. Three more orbital test flights followed, and after the last one President Ronald Reagan declared the shuttle fully operational. NASA, having promised cheap and reliable access to space, now had to fulfill that promise.

A launch schedule of twenty-four flights a year became, in the first years of shuttle operations, the overriding goal of the program. It was both a symbolic goal—like landing a man on the moon by the end of the 1960s—and an economic necessity. Low operating costs ("cheap") and a regular schedule ("reliable") both demanded frequent launches. NASA's promise to put payloads in orbit for $50 a pound (or even less) assumed that the shuttle fleet would make at least twenty-four fully loaded flights each year. Despite NASA's optimism, however, the goal of twenty-four flights a year proved elusive.

The job of preparing a just-landed orbiter for its next flight proved to be far more difficult, and far more time consuming, than NASA had anticipated. A 1972 report had predicted an average turnaround time of just under a week (160 hours) between flights, but a decade later the reality was closer to 10–12 weeks. In 1985, with four orbiters flying, NASA launched a total of nine shuttle missions. Four of the nine were flown by the orbiter *Discovery*, which averaged 8–9 weeks between landing and launch. It was an impressive record, but only if compared to the five- and six-flight years that had come before rather than the twenty-four-flight years that NASA still insisted were coming. It was also a record more impressive when viewed from the outside than from the inside. Technicians at the Kennedy Space

Center, where orbiters were "processed" between flights, were working at a headlong, unsustainable pace in the fall of 1985. Vital components were in short supply, and it became standard practice to "cannibalize" vital parts from an orbiter that had just landed and install them on another being prepared for launch. Worst of all, schedule-conscious NASA managers overlooked potentially serious safety concerns or gave low priority to finding solutions. Suggestions that NASA ought to slow down were unofficially but firmly discouraged.

"Go fever"—a single-minded determination to complete the mission—caught up with NASA on the morning of January 27, 1986. Seventy-three seconds after liftoff, the shuttle *Challenger* exploded. Hot gasses had burned through the zinc putty and rubber o-rings sealing the lowermost field joint in the left solid rocket booster. A jet of fire, spurting from the ruined joint like a cutting torch, burned away at the wall of the external fuel tank and at the metal strut that linked the bottom of the SRB to it. The strut broke, causing the still-firing SRB to pivot around its intact upper strut and rupture the external tank. Liquid hydrogen and oxygen spilled from the rupture and ignited almost instantly, fueling the explosion that blew the orbiter apart. The crew compartment, blown clear of the explosion, fell intact back to Earth. At least some of the seven crew members survived long enough to activate their personal oxygen tanks, but perished when the crew cabin slammed into the Atlantic two minutes later.

The *Challenger* disaster, and the investigation that followed, revealed a series of unpleasant truths about the shuttle program. Seals in SRB field joints had been damaged by hot gas "blow-by" on at least ten previous flights, but neither NASA nor Morton Thiokol (makers of the rocket motors for the SRBs) had thought the problem serious enough to suspend flights while it was solved. NASA had, in order to maintain its flight schedules, routinely granted "waivers" of safety checks required by its own operating procedures. Morton Thiokol's senior managers, under pressure from NASA officials, had overridden their own engineers' recommendation not to launch *Challenger* as scheduled. NASA officials had also, on the morning of the launch, brushed aside concerns that ice on the launch pad could damage the orbiter and that high seas off the Florida coast could make recovery of the SRBs impossible. The astronaut corps was routinely kept ignorant of engineers' concerns about the shuttle, and was not involved in deciding whether it was safe to launch *Challenger.* The presidential commission that investigated the disaster charged, in a scathing report, that NASA had abandoned its once-exemplary safety program in order to meet an overambitious schedule.

The first twenty-four flights of the shuttle (up to the loss of *Challenger*) showed that a reusable spacecraft was possible, just as the ten flights of

Project Gemini had shown that orbital rendezvous, space walks, and long duration flights—the basic elements of Project Apollo—were possible. A Gemini spacecraft, pushed to its technological limits, could have flown around the moon in 1966 or 1967, but only at great risk to its crew. It could not—as the later Apollo spacecraft did—have landed astronauts on the surface and returned them safely to the Earth. The shuttle, pushed to its operational limits, began to make space launches seem "routine" in late 1984 and 1985, but only (as the loss of *Challenger* revealed) at great risk to its crews. Post-*Challenger* changes in U.S. space policy and law reflected this. The shuttle would, according to the government's revised Space Shuttle Use Policy, be reserved for missions "that (i) require the presence of man, (ii) require the unique capabilities of the Space Shuttle, or (iii) when other compelling circumstances exist."

The shuttle fleet resumed operations in September 1988. *Endeavour*, a new orbiter designed to replace *Challenger*, arrived at Kennedy Space Center in May 1991 and made its first flight a year later, returning the fleet to its intended complement of four spacecraft. Talk of twenty-four flights a year was gone from NASA's public statements, however, and talk of cheap, reliable access to space was downplayed. The "official" image of the shuttle became that of an experimental vehicle used for scientific research and exotic jobs like in-flight repairs to the Hubble Space Telescope. That image would remain in place until February 1, 2003, when the orbiter *Columbia* broke up over Texas as it returned from a long, successful mission.

The accident investigation revealed that a chunk of foam insulation had broken off the external tank at liftoff, fatally damaging the heat-resistant tiles on the leading edge of *Columbia*'s left wing. It also revealed that midlevel NASA managers had expressed concerns about the possibility of damage from the foam strike, and been ignored. Fifteen years after the *Challenger* disaster, the shuttle's reputation as a fragile machine that NASA had pushed too hard returned with a vengeance. The three surviving orbiters were grounded and, at this writing, remain grounded.

THE ARIANE FAMILY

European nations had been in space since the early 1960s, buying space on American launch vehicles for satellites designed and operated by the multinational European Space Research Organization (ESRO). Europe had been working on a "native" launch vehicle for just as long. The European Launcher Development Initiative (ELDI) was formed in 1961 and spent the rest of the decade wrestling with a trouble-plagued three-stage launcher named Europa. France, meanwhile, was pursuing its own plans. Having already acquired intermediate-range ballistic missiles under the leadership of President Charles de Gaulle, it unveiled its own small launch vehicle, the three-stage *Diamant*, in 1965. The first *Diamant*, carrying a tiny satellite, lifted off from the Algerian desert—territory that France had agreed to vacate in 1966. The French shifted their launch operations, therefore, to the jungle town of Kourou in French Guiana. Kourou is, at first glance, an unlikely jumping-off point for outer space. A hot, damp backwater on the Caribbean coast of South America, it was separated by an ocean from France and by hundreds of miles from the nearest major city. Kourou had three critical advantages, however: open ocean to the north and east, a location near the equator that gave rockets a boost from Earth's rotation, and a legal status that made it (and the rest of French Guiana) part of France.

Frustrated by Europa's repeated failures and chafing under NASA restrictions on *what* they could launch on American boosters (no commercial satellites that would compete with those of U.S. firms), space-minded European nations changed their strategy. The once-separate ESRO and ELDI merged into the European Space Agency (ESA) in 1975, and the ESA quickly turned to France for help in developing a European booster. The result of the ESA-France collaboration, a new expendable launcher named Ariane, flew for the first time in 1979. A year later, France formed the Arianespace corporation to market the new booster's services.

Arianespace faced, in the 1980s, two principal competitors: NASA and American aerospace companies that were being encouraged by the Reagan administration to develop their own launcher-for-hire services. Throughout the decade, however, Arianespace enjoyed critical advantages over both. NASA anticipated that its new space shuttle would begin operations in 1978 or 1979, and that it would be able to launch payloads more cheaply than disposable boosters like Ariane. The shuttle, however, entered service later than expected and proved to be both less reliable and more expensive than hoped. The delays allowed Arianespace to establish itself in the market, and the shuttle's performance shortcomings allowed the Ariane to compete effectively with it. Arianespace shared one critical quality with NASA, however: government subsidies allowed it to sell its services for less than the cost of providing them. American companies—trying to start their own launch services with Atlas, Titan, and Delta boosters rendered "obsolete" by the shuttle—were all but shut out of the market. Forced to set launch costs high enough to cover operating expenses and make a modest profit, they found themselves regularly underbid by both NASA and the Europeans.

Arianespace's ability to offer high reliability and (artificially) low costs made it attractive to commercial firms throughout the Western world. The original Ariane gave way to the more powerful Ariane 2 in 1983, the Ariane 3 in 1984, and the Ariane 4 in 1989. All three launchers had a "core" similar to the original: first and second stages fueled with a self-igniting combination of nitrogen tetroxide and unsymmetrical dimethyl hydrazine, topped by a third stage fueled with liquid hydrogen and liquid oxygen. The Ariane 3 and Ariane 4 models, however, added strap-on solid-fuel boosters: two in the Ariane 3 and up to four in the Ariane 4. Increasing power gave the Ariane launchers increasing flexibility, including the ability to carry two satellites at once in a specially designed payload section. The 1986 *Challenger* disaster, and the subsequent U.S. banning of commercial payloads from the shuttle, made Arianespace the world's leading provider of commercial launch services—a distinction it has held ever since.

ENERGIA-BURAN

Aware that the United States was developing a reusable spacecraft and determined to have one like it, the Soviet Union authorized the Energia-Buran program in 1974. The design of the winged Buran orbiter blatantly copied that of the American space shuttle—a tacit admission by Soviet designers that they could not improve on the shuttle. Reports of the Buran in the Western press emphasized this, and many casual observers dismissed it as a mere copy. Such dismissals, however, overlooked a critical difference. Whereas the space shuttle was designed as a fully integrated system, the Buran orbiter was merely one of many payloads that could be carried by the new Energia booster being developed in parallel with it. Energia was, arguably, the technological core of the program: a Saturn-class launch vehicle with the flexibility of the long-serving Soyuz.

Energia reflected Soviet designers' preference for liquid-fuel rockets, and the sophisticated understanding of them that thirty years of experience had given them. It also reflected the designers' preference for modular launch vehicles that could be customized to meet the demands of particular missions. The basic Energia core stage contained four modular liquid-fuel engines fueled by liquid hydrogen and liquid oxygen (LOX). Two, four, or six strap-on boosters, each with a single LOX-kerosene engine, clustered around the core. Various upper stages, powered by LOX-kerosene engines of their own, could be mounted atop the core. Payloads could be carried atop the upper stage or, in the case of heavy or bulky items like Buran, carried alongside the core. The Energia was the most powerful launch vehicle ever flown—its payload of 200 tons exceeding even that of a Saturn V. Plans called for it to be the basis for a series of ambitious Soviet space projects that would unfold in the 1980s, 1990s, and beyond: laser-armed battle stations, missile defense systems, space telescopes, nuclear waste disposal, renewed exploration of the moon, and more.

In the end, Energia and the ambitious plans made for it came to nothing. The LOX-kerosene engines for the booster stage suffered from serious development problems, the Buran orbiter fell badly behind schedule, and the first launch of Energia in May 1987 was tarnished by the (unrelated) failure of the payload's guidance system. The second Energia test flight, which carried an unmanned Buran into orbit in October 1988, was a brilliant success. Energia itself performed flawlessly, and Buran completed two orbits and a perfect computer-controlled landing. The second flight was also, however, the functional end of the program. The Soviet Union had begun its rapid decline—its empire in Eastern Europe would collapse within a year—and there was no money left for experimental technology as expensive as Energia and Buran. Neither machine ever flew again, and the program outlived the Soviet Union by only two years before being dropped from the budget in 1993.

Had the Soviet Union not fallen, and had the program been allowed to proceed, Energia might have become the heavy-lift counterpart of Soyuz: a rugged, versatile workhorse built in large numbers and used for a variety of missions. The availability of such a launch vehicle might have made the Soviets' ambitious plans politically, as well as technologically, feasible. Its loss put the Soviet (later Russian) space program where NASA had been in the early 1970s: forced to reassess its future, and unable to be sure that it even had one.

10

Missiles after the Cold War, 1990–Present

The development of guided missiles was shaped, for forty-five years, by the Cold War: the rivalry of two nuclear-armed nations with half the world between them. Each superpower embraced nuclear-armed ICBMs and SLBMs as a means of deterring aggression by the other. Each also developed shorter ranged ballistic and cruise missiles that could carry nuclear warheads but were not tied to permanent bases. Wary of each other's long-range missiles, both superpowers experimented with ground-based and (in the case of the United States) space-based systems to defend against them. Wary of each other's long-range bombers, they developed surface-to-air and air-to-air missiles capable of destroying them. Planning for a full-scale European war that never came, they equipped their air and ground forces with lightweight guided missiles designed to destroy the other's aircraft, tanks, and fortifications. Over the course of the Cold War, the superpowers supplied missile systems to dozens of other nations. They remained, well into the 1980s, the world's leading sources of rocket and missile technology.

The breakup of the Soviet Union led Russia and the United States to rethink Cold War–era assumptions. The START II treaty, which banned heavy ICBMs and the use of MIRVs on ICBMs (though not SLBMs), was signed by Presidents George Bush and Boris Yeltsin in January 1993, and subsequently ratified by the U.S. Senate (1996) and the Russian Duma (2000). President Bill Clinton signed a separate agreement with Yeltsin in

1994, in which both nations agreed to retarget their ICBMs and SLBMs away from each other's territory. The agreement made little practical differences (the wartime target coordinates remained in the guidance system computers, and could be restored in a matter of seconds), but had immense symbolic value. It meant that, as Clinton said in a 1996 debate with Republican challenger Robert Dole, the United States and Russia no longer had missiles "pointed at each other."

These changes gave Americans and Russians alike reason for optimism. The end of the Cold War did not, however, mean an end to the development and deployment of missiles or to the threat that missiles posed to peace and security. Four decades in which the superpowers had passed missile technology to their allies could not be readily undone. Many of the world's most volatile regions—Israel, the Persian Gulf, the Korean Peninsula, and the Indian subcontinent—were awash in missiles sent there by the superpowers to advance their Cold War agendas. China, Israel, India, Pakistan, Iraq, Brazil, and other nations had developed the capability to copy American and Soviet missiles, manufacture their own missiles, and modify missiles acquired elsewhere. A thriving trade in missiles existed between these nations even before the Cold War ended. After the Cold War, it expanded and intensified.

The proliferation of missiles coincided with rising concerns about the proliferation of nuclear, biological, and chemical weapons—"weapons of mass destruction," or WMDs. Concerns about regional wars fought with nuclear missiles and the specter of guerillas and "rogue states" armed with missiles and WMDs replaced Cold War–era fears of a full-scale nuclear exchange between the superpowers. Missiles remained a critical threat to world peace and security in the 1990s and early twenty-first century, but in a disturbingly new and different way.

THE PERSIAN GULF WAR (1991)

Symbolically, the Cold War ended when cheering crowds pulled down the Berlin Wall in 1989. Politically, it ended when the Soviet Union dissolved itself in 1991. Militarily, however, it ended in January and February of 1991, when a coalition of Western and Arab nations routed the Iraqi army and rolled back Saddam Hussein's annexation of Kuwait.

The Gulf War was, in a sense, the last battle of the Cold War era: a massive land and air operation that pitted Coalition forces armed with American weapons against Iraqi forces armed with Soviet weapons. The Coalition's decisive victory, achieved in eight weeks with minimal casualties, vindicated a new generation of battlefield missiles developed during the last years of the Cold War to counter the Soviet Union. A-10 Warthog attack aircraft firing Maverick missiles and AH-64 Apache helicopters firing Hellfire missiles achieved success rates well above 95 percent against Iraqi tanks. Coalition fighters armed with the latest versions of the venerable Sidewinder and Sparrow missiles, as well as the new AMRAAM (Advanced Medium-Range Air-to-Air Missile), seized control of the sky on the first days of the war and never relinquished it.

The Gulf War was also, however, a sign of things to come. It saw the first operational use of the Tomahawk cruise missile, which became one of America's signature weapons in the low-intensity conflicts of the next decade. It also saw the bombardment of both Saudi Arabia and Israel with medium-range ballistic missiles, a strategy that made concerns about missile in the hands of "rogue nations" and guerillas seem prescient rather than alarmist.

The Tomahawk was designed to penetrate deep into enemy territory and destroy, with great accuracy, strategically valuable but heavily defended targets such as communications centers. Launched from surface ships or submarines, it flies at low altitudes and high (though still subsonic) speeds. The Tomahawks used in the Gulf War were steered by an inertial guidance system supplemented by a Terrain Contour Matching (TERCOM) system. TERCOM scanned the ground below the missile with a video camera, then used powerful, compact onboard computers to compare the video image to a specially prepared digitized "map" of the planned route. If the picture from the camera did not match the map, the system would then adjust the missile's course until it did. When accurately programmed, the TERCOM guidance system gave the Tomahawk unprecedented accuracy for a long-range missile: the ability, for example, to hit a specific building in the midst of a large city.

Coalition forces fired 278 Tomahawks during the Gulf War, nearly all of them from surface ships in the Persian Gulf. Their warheads—a 1,000-pound charge of high explosive or a dispenser that ejected hundreds of small "bomblets"—could not penetrate reinforced targets, but were devastating to ordinary buildings. Electric power stations, communications centers, and the headquarters of the Ministry of Defense all were among their principal targets. The success rate for cruise missiles fired in the first days of the war was exemplary. The first wave of Tomahawks hit their assigned targets 85 percent of the time, as did 31 of 35 (89 percent) of the ALCMs launched by B-52 bombers on the first night of the war. Ironically, later waves of Tomahawks became less accurate because of the success of the air campaign. The maps in their TERCOM systems could not be updated fast enough to compensate for damage done to roads, bridges, and other landmarks by earlier attacks.

The Tomahawk cruise missile represented the cutting edge of missile technology. The Scud B ballistic missile was, in contrast, an obsolete relic of the late 1950s that the Soviet Union (which had supplied them to Iraq) had long since retired. The Al-Hussein missile was an Iraqi-made derivative of the Scud B that offered longer range (350–400 rather than 180 miles) but no better accuracy. They could hit a given city, but not a particular neighborhood—much less a particular building—within that city. The two missiles were, nevertheless, Iraq's strategic "ace in the hole." Unimpressive as they were, they were still ballistic missiles, and so were almost invulnerable once they left their launchers. The Iraqi Air Force lost, in the first days of the war, any ability it might have had to strike at targets beyond Iraq's borders. The Scud and Al-Hussein missiles had that ability, and their ability to carry biological and chemical warheads made them a serious threat despite their lack of accuracy. The missiles' value as a weapon of intimidation had been established a few years earlier, in the Iran-Iraq War of 1980–1988. Iraq's bombardment of Tehran and other Iranian cities in 1987–1988 had killed 2,000, injured 6,000, and displaced thousands more who had fled the cities fearing that the *next* missile might carry the poison gas that Iraq had already used against Iranian troops. More significant, from Iraq's point of view, the missile bombardment had brought the Iranians to the negotiating table and forced them to accept unfavorable peace terms in order to end the war quickly.

Saddam Hussein intended the 1991 missile attacks on Israel and Saudi Arabia to give him similar political leverage against the Coalition. The sixty missiles aimed at Saudi Arabia targeted areas where American troops were quartered. They were likely intended to cause mass casualties, leading to an American withdrawal from the Gulf as the truck bombing of the U.S. Marine barracks in Beirut had led to the withdrawal of troops from Lebanon in 1983. The sixty missiles aimed at Israel targeted major cities and were clearly intended to draw Israel into the war. American leaders believed, with reason, that Israeli involvement in the war would cause Saudi Arabia and other Arab states to withdraw from the Coalition and close ranks with Iraq. Iraqi, also with reason, evidently believed it, too.

The 1991 missile bombardment, which began on the night of January 17–18 and continued sporadically into late February, failed to achieve its goals. Israel, making a historic exception to its long-standing policy of meeting force with force, did not retaliate against Iraq. The Coalition remained intact, and the missile attacks had no real effect on the pace of the war. Casualties were astonishingly light. A single Scud hit a U.S. Army barracks in Al Khobar, Saudi Arabia, killing twenty-eight soldiers and injuring ninety-eight on the night of February 25. Thirty-nine missiles fired into

Israel over the course of five weeks caused only two deaths and eleven serious injuries.

The bombardment failed for a variety of reasons. The first, and most significant, was the sheer inaccuracy of the missiles themselves. The Scud attacks failed to achieve these goals for several reasons. One was the sheer clumsiness of the missile itself. The many that broke up in flight or fell harmlessly in the desert made the few that actually hit and did damage seem like isolated potshots rather than part of an overwhelming attack. A second reason was Iraq's misreading of the Israelis, who (like the British in World War II) had carried government-issued gas masks and camped out in improvised shelters to wait out the attacks. A third reason was behind-the-scenes American appeals to Israel to stay out of the war, coupled with American guarantees of protection. A fourth reason was a concerted campaign by Coalition aircraft and ground-based commando teams to locate and destroy Scud launchers. The fifth and final reason was the U.S.-built Patriot missile system, which was used to destroy incoming Scuds in flight.

A ground-based, radar-guided missile, the Patriot had been designed in the late 1960s to destroy Soviet bombers. Originally called the SAM-D (Surface-to-Air Missile Development), it was renamed in 1976 in honor of the nation's bicentennial. The Patriot, like other radar-guided SAM systems, consisted of missiles and ground-based radars linked by a control center. What made it revolutionary was the missile's ability to read, interpret, and respond directly to signals from its onboard radar. Patriot missiles could, in other words, make in-flight adjustments to their course without being "told" to do so by ground-based computers. The guidance system made the Patriot nimble enough to hit supersonic bombers and, in principle, even ballistic missiles. It was developed by the Army—and sold to a skeptical late-1970s Congress—primarily as an antiaircraft missile, but the possibility of using it as an antiballistic missile remained. A Patriot successfully destroyed a missile for the first time in a 1986 test, and the following year an upgraded model (the PAC-2) did the same. Part of the buildup for the Gulf War was a crash program by the Army and the Raytheon Corporation to produce PAC-2s and deploy them to the Gulf as antiballistic missiles.

Wartime reports of the Patriot's success were nothing short of spectacular. A Patriot battery in Saudi Arabia was credited with destroying an incoming Scud on the first night of the war. Six Patriot batteries deployed along Israel's eastern border were credited with destroying or deflecting dozens of Scuds that would otherwise have fallen on Haifa and Tel Aviv. Like the precision-guided "smart bombs" said to be capable of hitting specific doors or air shafts on a target building, Patriots emerged as one of the mechanical "heroes" of the war. The Patriot's image became that of "the little missile that could," and commentators compared its destruction of incoming Scuds to the fairy-tale heroics of Jack the Giant-killer. General H. Norman Schwartzkopf, commander-in-chief of the Coalition's military forces, announced at one point that the Patriot's success rate was perfect: thirty-three Scuds engaged, thirty-three destroyed or rendered harmless.

The reality, as postwar studies made clear, was more complex. Patriot crews discovered, early in the war, that the launch control system often failed to distinguish between incoming missiles and electronic "clutter." The opening-night Scud "kill" in Saudi Arabia was written off as such an error in 1992. The problem was particularly acute in Israel, where the Patriot batteries were sited near airports, cities, and other sources of stray electronic signals. The tendency of Scud and especially Al-Hussein missiles to break up in flight created more problems. Used in "automatic" mode, the Patriot's launch system often targeted not only the lethal warhead but also the largest chunks in the cloud of metal fragments trailing behind it. Programmed to launch two Patriots at each incoming missile but reading each fragment as a separate target, the launch computers sometimes sent as many as ten Patriots after a single disintegrated Scud. The Patriot's onboard radar could also be confused by these inadvertent "decoys." Postwar studies showed that as few as 9 percent of Patriots fired actually destroyed incoming missile warheads.

The Gulf War taught three critical lessons about the role of missiles in the post–Cold War era. The first was that cruise missiles like the Tomahawk could be a powerful weapon for precision attacks in densely populated areas. The second was that even crude ballistic missiles represented a significant threat; had the Scuds carried chemical warheads, the death toll from their attacks would have been higher and the political repercussions greater. The third lesson was that there was no single, easy way to eliminate the threat that such missiles posed.

MISSILES, ROCKETS, AND LIMITED WAR

The United States repeatedly intervened in foreign conflicts in the decade between the end of the Gulf War in February 1991 and the beginning of the "War on Terrorism" in October 2001. The interventions centered on Iraq but spread over three continents, encompassing Bosnia in southeastern Europe, Afghanistan in south central Asia, and the Sudan in northeastern Africa. Their stated objectives ranged from the containment of Saddam Hussein and the checking of Serbian nationalist leader Slobodan Milosevic to the persecution of the al Qaeda terrorist network after 1998 bomb attacks on U.S. embassies in Dar es Salaam, Tanzania, and Nairobi, Kenya. All the interventions had two essential qualities in common: they were limited in scope and duration, and they hinged on the use of advanced weapons—especially air power.

The military operations of 1991–2001 were shaped by the "Powell Doctrine," named for General Colin Powell, chairman of the Joint Chiefs of Staff during the Gulf War. Formulated in response to the long, indecisive war in Vietnam, it stated that the United States should intervene only when it could bring overwhelming force to bear on the enemy. The Gulf War, which took place during Powell's term as chairman, was widely seen as a vindication of the doctrine. The ill-fated 1993 effort to capture Somali warlord Mohammed Farah Aidid, which resulted in the deaths of eighteen American soldiers in Mogadishu, was taken as proof that deviating from the doctrine would bring disaster. Cruise missiles—weapons that destroyed targets efficiently but put no American lives at risk—fitted the demands of the Powell Doctrine well. They became, therefore, a central part of the "small wars" that the United States fought in the 1990s.

Operations in Iraq, designed to enforce the terms of the Gulf War peace agreement, dominated American military operations between 1991 and 2001. On January 17, 1993, for example, U.S. forces fired forty-six Tomahawk cruise missiles at a complex outside of Baghdad that was suspected of housing a covert nuclear weapons program. Forty-two of the missiles launched successfully, and thirty-four hit the target, destroying eight buildings. On June 26 of the same year, another twenty-three cruise missiles struck targets in Baghdad, including Iraqi intelligence headquarters. Operation Desert Strike, a U.S. response to Iraqi troop movements threatening the Kurds of Northern Iraq, included a total of forty-four cruise missiles fired on September 3–4, 1993, against Iraqi air defense sites. Hundreds more were launched on December 16–19, 1998, as part of Operation Desert Fox, a systematic attempt to "degrade" the Iraqi military after the expulsion of UN weapons inspectors.

Cruise missile operations in the 1990s were not confined solely to Iraq. When the United States intervened in Bosnia as part of a NATO peacekeeping force, they initially relied solely on piloted aircraft to carry out missions against Serbian and Bosnian Serb forces. Beginning in mid-September of 1995, however, NATO and U.S. commanders agreed to use Tomahawk cruise missiles against Serbian air defense sites in and around Bosnian cities such as Banja Luka. The missile strikes were designed to make the skies over Bosnia safer for NATO aircraft, allowing attacks on Serb ground targets to continue uninterrupted. The cruise missile attacks that followed the embassy bombings of September 1998 were, in contrast, intended purely as a form of retaliation. The missiles fired into Afghanistan targeted suspected al Qaeda training camps along the Pakistani border, and those fired into southern Somalia were aimed at a pharmaceutical factory suspected of producing biological weapons. When operations against al Qaeda spread to Afghanistan in October 2001, cruise missiles once again played a crucial role. Their ability to hit specific targets with precision proved to be as useful in the (relatively) sparsely populated country as it had in densely populated Baghdad (see Figures 10.1a and 10.1b).

President Bill Clinton, who authorized all but the last of the cruise missile attacks listed above, was roundly criticized for them by his political opponents. Republican Senator Frank Murkowski of Alaska referred in a 1999 floor speech to "President Clinton's propensity to fire off cruise missiles apparently on a whim," and conservative talk-radio hosts such as Rush Limbaugh pointedly compared the number of cruise missiles expended by the Clinton administration with the significantly smaller number used by the Bush administration during the Gulf War. Cruise missiles became, for Clinton's detractors, evidence of what they saw as his general weakness on defense issues and his unwillingness to wage "real war" as President Bush had.

The use of cruise missiles and other precision-guided weapons by the United States also drew criticism from the political left. Peace activists, including some surgeons, particularly objected to the military's use of the term "surgical" to describe air raids, and its implied equation of destruction with healing. The suggestion that surgery could be performed with explosives was, they contended, a linguistic distortion worthy of "doublespeak"— the intentionally misleading language used by the totalitarian government of George Orwell's novel *1984*.

Cruise missiles offered, both in the Gulf War and in subsequent operations, a way of destroying targets efficiently and precisely. The government of Israel, fighting a more intimate kind of war with Palestinian guerillas in the West Bank and Gaza Strip, used rocket-firing helicopters to achieve the same goals. Helicopters could move quickly over the crowded streets of Palestinian towns and cities, zeroing in on the source of guerilla strongholds while keeping Israeli personnel safely isolated. Unguided rockets, fired at point-blank range, were more than capable of destroying buildings or cars linked to the guerillas—a standard form of Israeli retaliation. They could also be used as weapons of assassination against known or suspected guerilla leaders caught in the open. On March 24, 2004, for instance, an Israeli helicopter fired three missiles at Sheikh Ahmed Yassin as he left a mosque in Gaza City, killing him instantly. Israeli officials characterized Yassin, a leading member of the Palestinian militant group Hamas, as a notorious terrorist. Many outside observers, however, shared the Palestinians' outrage. They argued that firing missiles directly at an unarmed, unprotected man on a city street crossed the line separating military expediency from senseless brutality.

Palestinian guerillas have also made extensive use of rockets and rocket-propelled grenades (RPGs) in their ongoing struggle with the Israeli military. It was in Iraq, however, that RPGs have come into their own as a weapon for low-intensity urban combat. The RPG has become the signature weapon of the Iraqi guerilla fighters who have taken a steady toll on American soldiers and civilian contractors in 2003–2004. Portable, concealable, and easy to operate, RPGs are nonetheless powerful enough to destroy cars, trucks, and other unarmored vehicles. The U.S. military's ubiquitous "Humvee" utility vehicles have proved especially vulnerable to RPG fire, and the lack of adequate armor protection for their crews has prompted sharp criticism of the Defense Department's prewar planning. In April and May 2004, Garry Trudeau's long-running cartoon strip *Doonesbury* included a storyline in which one of the main characters loses a leg when his unit is attacked (offstage) by Iraqi guerillas. Later, in the hospital, a fellow soldier asks him how he was wounded. He replies, simply: "Ambushed, outside Fallujah. My humvee took an RPG." Trudeau assumes that his readers need no further explanation, and no definition is given of what would once have been an obscure military acronym. The fact that he can do so suggests how large a role rocket-propelled weapons have come to play in the ongoing conflict in Iraq.

MISSILE PROLIFERATION AND MISSILE CONTROL

The proliferation of missiles in the developing nations of the world began while the Cold War was still going on. Both superpowers made short- and medium-range missiles available to their European allies, and both superpowers (the United States occasionally and the USSR extensively) supplied missiles to more distant and less technologically advanced allies. European nations such as France and Germany also contributed to proliferation, supplying missiles, missile components, and "dual-purpose" components that could be used for civilian purposes or adapted for use as missile parts.

Over time, missile proliferation gradually took on a life of its own. China, one of the earliest recipients of Soviet missile technology, developed its own ballistic missiles by copying Soviet designs. It subsequently became a major supplier in its own right, selling its Soviet-derived missiles throughout the developing world. A number of countries—notably North Korea, Pakistan, Iran, and Iraq—followed China's lead. They began by acquiring missiles from others and moved on to modifying, copying, and producing missiles of their own. Others—such as Israel, India, and Argentina—

Figures 10.1a and 10.1b: A two-sided pictorial leaflet used during Operation Enduring Freedom (2001–2002) threatens Taliban soldiers in Afghanistan with the power of American precision-guided missiles. In the first image, a group of armed men spots a missile heading for the cave in which they are hiding. In the second, the missile has struck: Two men are buried beneath the rubble, and the rest are trapped. The inscription reads "(front) Taliban, do you think you are safe . . . (back) in your tomb?" Courtesy of the Department of Defense and the Library of Congress, image numbers LC-DIG-ppmsca-02031 and -02032.

149

developed missiles purely for their own use. Israel had its Jericho, India its Agni, and Argentina its Condor. Even Brazil, as far removed from the Cold War arms race as any major nation on Earth, began to produce a launch vehicle, the Sonda, that could be converted into a missile. Unlike its neighbor Argentina, it became an enthusiastic exporter of both missiles and missile components. The end result was a global trade in missiles that completely bypassed the United States, the USSR, and the rest of the world's leading industrialized nations. The missiles fired in the Iran-Iraq war of 1980–1988 were built in Iraq, Iran, China, North Korea, and Libya.

Several of the developing countries that built, sold, or sought to acquire missiles had a history of political instability. Others were virtual dictatorships, run by military leaders with a history of aggression toward neighboring states and violence toward their own citizens. Still others were parties to decades-old territorial disputes. Many of the most active missile-building and -buying states in the developing world possessed, or were attempting to acquire, weapons of mass destruction. The prospect of aggressive leaders in politically unstable countries armed with missiles carrying biological, chemical, or nuclear warheads left the leaders of the United States, the Soviet Union, and most of Europe profoundly uneasy. Out of that unease came, in the late 1980s, the Missile Technology Control Regime (MTCR).

The MTCR is an informal partnership whose members agreed to limit trade in missiles and missile components. It was originally formed, in 1987, by the United States, Canada, Britain, France, Germany, Italy, and Japan. By 1993, it had expanded to include all the major nations of Western and Central Europe as well as Australia, New Zealand, and Argentina. Brazil, South Africa, and Russia joined in 1995; Poland, the Czech Republic, and the Ukraine in 1998; and South Korea, the last addition to date, in 2001. The guidelines to which members of the MTCR agree to adhere state that they will strictly limit their exports of complete missiles, missile subsystems (warheads, guidance systems, rocket motors), plans and designs for missiles, and equipment for manufacturing missiles. The guidelines also establish a second category of materials and equipment that, although they have legitimate peaceful uses, could be adapted for use in missiles. Members of the MTCR have greater freedom to export "Category 2" items, but are still expected to choose their trading partners judiciously.

"Missile" is defined broadly in the MTCR guidelines. It includes ballistic missiles, cruise missiles, sounding rockets, space launch vehicles, remote-controlled aircraft—any unpiloted machine capable of carrying a 50-kilogram (110-pound) payload for 300 kilometers (180 miles). The Scud B and anything more powerful falls under the guideline, as do most space launch vehicles. The guidelines explicitly allow the transfer of technology designed to support national space programs, but recognize that the line between space launch vehicles and ballistic missiles is very thin. Members of the MTCR that export space launch technology pledge, therefore, to obtain and monitor the importer's assurances that it will be used only for peaceful purposes.

The MTCR was (and is) a significant barrier to the proliferation of weapons of mass destruction and missiles capable of carrying them. Russia is no longer selling ballistic missile systems abroad (as the Soviet Union did throughout the Cold War), and technologically advanced countries such as France and Germany are no longer freely exporting guidance systems and other critical technologies (as German companies did to Iraq in the late 1980s and early 1990s). At least two member nations—Argentina and South Africa—have shut down their ballistic missile programs entirely.

Despite these significant achievements, the MTCR has major limitations. One in particular stands out: few of the developing nations most closely associated with missile proliferation are members. China, Israel, India, Pakistan, and North Korea, for example, are all conspicuously absent. All five possess short- and medium-range ballistic missiles as well as nuclear weapons, and all except Israel worked to develop longer range missiles during the 1990s. China, for example, spent the decade test-flying its DF-31: a three-stage, solid-propellant ICBM. Capable of carrying a 1-megaton nuclear warhead (or three smaller warheads) over 8,000 kilometers, it can reach the Pacific Northwest from launch sites in Manchuria. India suspended development of its Agni IRBM in 1995, but under the newly elected government of A. B. Vajpayee decided to begin development of an improved Agni II (with a range of 3,000 km) in March 1998 and a still-more-advanced Agni III (with a range of 3,500 km) in early 1999. Both later models of the Agni would have the ability to strike targets in China (as well as nearby Pakistan) with nuclear warheads. North Korea's Nodong 1, based on the basic Scud design but with a range up to 1,500 kilometers, first flew in 1993 and has since been built and deployed not only in North Korea but also in Pakistan (as the Ghauri II) and in Iran (as the Shehab 3). It also formed the first stage of the two-stage Taepo-Dong missile, which first flew in 1998 and was exported to Iran as the Shehab 4.

Iraq was a special case throughout the 1990s. United Nations Security Council resolution 687, which established the cease-fire terms that ended the Gulf War, prohibited Iraq from possessing missiles (or components or support equipment for missiles) with ranges greater than 150 kilometers. It provided for UN inspection teams to monitor compliance with the resolution by overseeing the destruction of prohibited missiles. UN inspectors led by Rolf Ekeus supervised the destruction of forty-eight missiles in July 1991. Iraq claimed to have destroyed the balance of its arsenal without UN oversight between July and October of that year, but admitted in March 1992 that it had, in fact, hidden eighty-five of those missiles, along with warheads and launchers. The final disarmament of Iraq, carried out over the next dozen years, was too complex a process to recount here. It involved further UN resolutions, disclosures, deceptions, punitive air strikes, complex negotiations over what the inspectors could inspect, and finally—in the fall of 1998—Iraq's expulsion of the UN inspectors.

The issue of Iraqi missiles was bound up, from 1991 on, with that of Iraqi weapons of mass destruction. A congressional task force on terrorism reported, six months before UN inspectors were driven out, that forty-five Scud missiles with biological or chemical warheads still existed inside Iraq, and that other Iraqi missiles (some with ranges of up to 3,000 kilometers) had been hidden in Libya and the Sudan. President George W. Bush stated flatly—in a speech delivered in Cincinnati on October 7, 2002—that Iraq possessed missiles capable of striking Israel, Saudi Arabia, or Turkey, and threatening more than 130,000 Americans living and working in the Middle East. He alluded, in the same speech, to the possibility that Iraq was developing remote-controlled aircraft capable of dispersing biological weapons over the United States. The Bush administration's conviction that Iraq possessed both missiles and weapons of mass destruction, and was thus a significant danger to the United States and the world, set the stage for the U.S.-led invasion of Iraq in March 2003.

MISSILE DEFENSE

The worldwide proliferation of missiles in the 1990s led the United States to reconsider, as the decade ended, the idea of building a national missile defense system. Unlike the Strategic Defense Initiative of the 1980s, the National Missile Defense (NMD) system was not designed to absorb a full-scale Soviet or Russian first strike. It was, rather, envisioned as a defense against a more limited attack: a handful of missiles launched by a "rogue state" such as North Korea or Iraq. President Bill Clinton, commenting on NMD in March 1999, referred explicitly to the possibility that such states could use missiles to deliver weapons of mass destruction. Central Intelligence Agency director George Tenet, addressing the Senate Foreign Relations Committee in March 2000, specifically cited North Korea, Iran, and Iraq as sources of ballistic missile threats over the next fifteen years.

The NMD system, as envisioned in the late 1990s, consisted of ground- and space-based radars to track incoming missiles and ground-based interceptor missiles to destroy them. It was, in its broad conception though not in its engineering details, closely related to the Patriot missiles deployed during the Gulf War. The Patriot, however, made its interceptions relatively close to the ground. The NMD interceptors were designed to destroy incoming warheads while they were still outside Earth's atmosphere. Doing so successfully would require the system to distinguish the real warhead from decoys and other distractions, and hit it at high speed and high altitude. The Department of Defense authorized development work on the system in 1996. The original schedule called for a total of nineteen flight tests of the system between 1997 and 2005, with the critical decision on whether or not to continue the program made after the sixth test in mid-2000. Only five of the six tests were completed by the deadline, and the results were ambiguous at best. The first two tests were not designed to produce an interception. The next three, which were, resulted in one success and two failures, but the second failure was due to problems with the interceptor's off-the-shelf rocket motor rather than a specially designed component like the guidance system.

President Bill Clinton chose, on September 1, 2000, to defer a final decision on NMD to his successor. The results of the tests probably contributed to his decision, as did the knowledge that he would be leaving office within the year. The opinions of other world leaders also played a role in his decision. Russia and China firmly opposed the system, fearing that it would allow the United States to threaten them with its own missiles with no fear of retaliation. Western European countries also had reservations about the system, fearing that it would lead to a reduced American commitment to NATO and to the erosion of U.S.-Russian arms control treaties. Prominent members of Clinton's own party—former Secretary of Defense William Perry and former Senator Sam Nunn, for example—encouraged the president to defer his decision.

The decision on the future of NMD was made, therefore, by President George W. Bush, who had already signaled his interest in and commitment to it while a candidate. Bush announced, on December 13, 2001, that he intended to withdraw the United States from the ABM Treaty of 1972, which would have barred deployment of a system like NMD. Six months later, in June 2002, the withdrawal took effect. Six months after that, on December 17, 2002, he announced his decision to deploy the first elements of the NMD system in 2004 and 2005. It was, he argued, a response to "perhaps the gravest danger of all: the catastrophic harm that may result from hostile states or terrorist groups armed with weapons of mass destruction and the means to deliver them."

There is no question that nations actively hostile to the United States now possess both weapons of mass destruction and ballistic missiles. Other questions, however, remain open: whether such nations will choose to develop missiles capable of striking the United States, whether they would use such missiles in anger, and whether ground-based interceptors would be capable of stopping such an attack.

11

Conclusion: What Next?

◆

Technological change is too complex a process to sum up with the old maxim that "necessity is the mother of invention." New technologies rarely spring into existence simply because their inventors see an unmet need. Even so, changes in an existing technology often result from changes in the demands that are placed upon it. So it is likely to be with rockets in the twenty-first century.

The demands placed on the rocket motors that drive military missiles are well-established, and have been for decades. Missiles have evolved in those decades, but the major changes in their design have involved guidance systems, reentry vehicles, and warheads—not propulsion systems. Missile propulsion systems have been improved in recent decades, but the improvements have been incremental rather than systemic. Existing rocket motors produce more than enough thrust to lift the largest military payloads. They are highly reliable, easy to ignite, and well-suited to producing intense thrust for short periods of time. Their principal shortcoming—the speed at which they wear out—is irrelevant since they are designed for a single mission that invariably ends in their destruction. They are, from a military standpoint, a well-tested technology that meets all foreseeable needs, and a decision to scrap them in favor of something radically new and untested seems unlikely.

The same appears to be true of space launch vehicles. Russia is still launching payloads and cosmonauts atop its venerable Soyuz booster, and has now licensed a version of it to the European Space Agency as a supplement to its homegrown Ariane family of boosters. The "Long March" booster that carried Chinese "taikonaut" Yang Liwei into orbit on October 16, 2003, is part of the same Soyuz lineage. NASA's space shuttle, the last great breakthrough in launch vehicle technology, is slated for retirement by 2010. Its replacement, not yet chosen, is likely to be a new disposable launch vehicle capable of lifting cargo or a small, winged, reusable spacecraft designed solely for transporting passengers. Burt Rutan's *Spaceship One*, a new first step toward the old dream of a low-cost reusable spacecraft, is radical in concept but uses relatively conventional jet and rocket engines. Any spacecraft capable of winning the "X Prize"—$10 million to the builders of the first spacecraft that can make two manned suborbital flights within fourteen days before 2005—will almost certainly do the same.

The third major use for rocket engines is the propulsion of spacecraft in (rather than into) space. It is there—outside Earth's atmosphere—that the demands on rocket engines are changing radically, and there that radical changes in rocket technology are likely to follow.

President George W. Bush called, in a speech on January 14, 2004, for the United States to shift the focus of its space program away from operations in Earth orbit and toward the exploration of other worlds. He called, first, for a return to the moon and for the establishment of a permanent base there. Going back to the moon will require, for the first time in thirty years, manned spacecraft designed to operate for extended periods in deep space. The spacecraft will naturally draw on Apollo-era designs, but if the base is to be permanently staffed it is difficult to imagine that the spacecraft built to service it will be designed only for a single flight. The high cost of disposable spacecraft could be justified in the context of the Apollo program because of the late President Kennedy's looming end-of-decade deadline. It was also made palatable by political leaders' conviction that *going* to the moon—not exploring it—was the point of the program, and that there would be relatively few flights. Establishing and staffing a permanent base would be a long-term program in which flights to and from the moon were only a means to an end. It would, almost certainly, demand a vehicle like the C-130 transport aircraft that service scientific stations in Antarctica: powerful, versatile, and utterly reliable.

The single most significant design choice made by Apollo-era mission planners was to land on the moon with a spacecraft designed to operate solely in airless space. The designers of the Apollo LM took full advantage of the fact that their spacecraft did not need to penetrate Earth's atmosphere or land in full terrestrial gravity. The advantages of a specialized space-only ship have not diminished since the last LM left the moon in

December 1972. A spacecraft designed to serve a permanent lunar base would very likely take advantage of them. Building a *reusable* space-only ship would, however, mean overcoming a major technological challenge: refueling and, if necessary, refurbishing or replacing its engines "in the field." Meeting that challenge—through different propellants, more efficient engines, or other innovations—is likely to produce a propulsion system notably different than the one that drove the LM.

The second major element of President Bush's agenda for spaceflight—landing humans on Mars—poses even greater technological challenges and, if carried out, is likely to demand even greater levels of innovation. The moon is roughly a quarter-million miles from Earth: three days' journey at Apollo-era speeds. Mars, even when relatively close to the Earth, is well over 33 million miles away: a journey of eight or nine months at minimum. An expedition to Mars involving a human crew would likely require both a crew larger than that used on the Apollo missions and a ship larger, relative to its occupants, than the Apollo spacecraft. The engines for such a ship would have to be larger than the ones used for the Apollo spacecraft, and capable of firing after weeks or months (not just days) of soaking in the cold of space. They would also, given the length of the trip, require massive quantities of propellants.

One of the most promising plans for sending humans to Mars, popularized by Dr. Robert Zubrin and alluded to by President Bush in his 2003 speech, proposes that the propellant problem be solved by tapping resources available on Mars. The human crew that lands on Mars would, in this plan, be preceded by the ship that will bring them home. Launched two years before the manned ship (optimum conditions for an Earth-to-Mars launch occur every twenty-six months), the return ship would be powered by rocket engines burning methane (CH_4) and oxygen. It would land, unmanned, on the Martian surface and deploy a small nuclear fission reactor. Then, using a chemical catalyst and electricity from the reactor, it would use carbon dioxide (CO_2) from the Martian atmosphere and "imported" hydrogen from Earth to manufacture methane and water. The methane would be used to replenish the ship's fuel tanks, and the water broken down, by electrolysis, into hydrogen (used to make more methane) and oxygen for the ship's oxidizer tanks. Additional oxygen could, if necessary, be obtained by breaking down Martian CO_2, and venting the carbon monoxide (CO) by-product into the atmosphere. The human crew would leave for Mars knowing that their "return vehicle" would be waiting for them, fully fueled, when they arrive.

Methane-oxygen rocket engines would be a departure from existing practice, and building a spacecraft that could refuel *itself* on the surface of a distant planet would be a daunting challenge. Still, as Zubrin notes, the chemistry involved has been used in industry since the nineteenth century, and 100-kilowatt nuclear reactors have been in use since the mid-1950s. Assuming that the technological challenge can be met, the operational benefits would be enormous. The plan that Zubrin promotes, and that President Bush implicitly endorsed, would eliminate the need for any spacecraft to carry fuel for a two-way trip. Cutting the fuel load in half would reduce the mass and bulk of the ship by eliminating the need to haul nine months' worth of fuel and fuel tanks to Mars as dead weight. Whether or not self-refueling ships with methane-oxygen rocket engines turn out to be the method chosen to send humans to Mars, they represent a radically different way of thinking about propulsion systems for long-distance space flights.

An even more innovative propulsion system for such flights has already been tested, on a small scale, by a robot probe named Deep Space 1. Launched in 1998 to study Comet Borrelly, Deep Space 1 was also designed as a technology demonstrator: a tool for evaluating new design features that might be used on later spacecraft. One of those features was an ion engine, or electric thruster: a rocket that accelerates without burning its fuel.

Most atoms have a neutral electric charge; an ion is just an atom with a positive or negative electric charge. Like electrical charges repel one another, so a negatively charged ion placed between two negatively charged plates will be shot out from between them like a watermelon seed squeezed between two fingertips. An ion engine uses a small nuclear reactor to create a stream of ionized gas (that is, a gas whose atoms have been electrically charged) and accelerate it to fantastic velocities by passing it between electrically charged plates. Instead of the massive, short-term acceleration delivered by a chemical rocket, ion engines provide a slow, gentle, continuous acceleration. Their key advantages are simplicity, long life, and extraordinary fuel efficiency. A backup version of the ion engine from Deep Space 1 ran continuously for three and a half years in a test stand at NASA's Jet Propulsion Laboratory. Over its lifetime, an ion engine can deliver twenty times more thrust per pound of fuel than a traditional chemical rocket.

Beginning in 2006, a small ion engine like the one from Deep Space 1 is slated to drive the Dawn spacecraft on a mission to the large asteroids Ceres and Vesta. Carrying a little over 200 kilograms (440 pounds) of propellant, it will reach its destination far more quickly and efficiently than it could have with chemical rockets. NASA also has plans for a larger ion engine, dubbed Nexis for "nuclear electric xenon ion system," that would produce ten times as much electricity (and thus significantly more thrust) than its smaller cousin on Deep Space 1 and Dawn. Tested at JPL in December 2003, the Nexis engine is designed to run for ten years (as opposed to two or three for the smaller version) and move 2 metric tons (2,000 kilo-grams) of propellant between its plates in that time. It is being considered for use in the 2009 Mars Smart Lander mission (which includes a robot capable of returning samples to Earth) and in still-in-development missions to three of Jupiter's four largest moons. Both missions would be well-served by high thrust-to-weight ratios—exactly what ion engines have the potential to provide.

Rockets have existed for close to a thousand years. Only in the last hundred years, however, have they evolved from pasteboard tubes filled with black powder to ion thrusters streaming glowing blue gas. The rockets designed, built, and flown in that hundred years bear the stamp of the societies in which they were created. They were shaped not only by imaginations of their designers, but also by political, economic, social, and cultural forces well beyond the designers' control. It is equally true, however, that the rockets of the last hundred years have left their mark on human societies: not only on individual nations but also on the ways in which nations deal with one another. Before the 1950s, we lived in a world where rockets existed. Since then, we have lived in a world that rockets played a large role in creating.

Glossary

Every effort has been made, in this book, to avoid technical terms except when they are absolutely necessary to tell the story. The technical terms that fall into this category, and so appear repeatedly, are defined here.

Antiballistic missile (ABM). A missile designed to intercept and destroy ballistic missiles or their warheads in flight. Antiballistic missiles are one form of missile defense system and, at this writing, the only form to be tested under real-world conditions.

Ballistic missile. A missile that travels in a long, parabolic trajectory, arcing upward under power and then falling back through the atmosphere to strike its intended target.

Black powder. A mixture of charcoal, sulfur, and saltpeter (potassium nitrate) similar, but not identical, to gunpowder. Black powder was the standard rocket propellant from the Middle Ages to the early twentieth century.

Booster. Technically, a self-contained rocket motor designed to be attached to a rocket or missile to give it extra power at liftoff. It is also used, casually, to refer to a launch vehicle, as in "Atlas Booster" or "Soyuz Booster."

CEP. Acronym for "circular error probable," a measure of a missile's accuracy. The CEP of a missile is the radius of a circle, centered on a given target, within which 50 percent of the missiles fired at that target will land.

Cold launch. A launch technique, used aboard ships and in some land-based missile silos, that ejects the missile from its storage container with a charge of compressed gas. The missile's own motor ignites only when it is well clear of its container. Compare to "hot launch."

Combustion chamber. The part of a liquid-propellant rocket engine in which fuel and oxidizer are combined and burned.

Cruise missile. A guided missile shaped like a small airplane. A typical cruise missile has some form of wings and tail, and uses a jet engine as its primary propulsion system (though it may also have a booster rocket for takeoffs).

Fuel. One of two principal components of a rocket's propellant (the other is oxidizer). Early liquid-propellant rockets used fuels such as gasoline, kerosene, and alcohol. Early solid fuel rockets of the 1940s used asphalt and similar compounds. "Fuel" is sometimes used, casually, as a synonym for "propellant," as in "liquid-fuel and solid-fuel rockets."

Guidance system. The electrical and/or mechanical system that steers a missile while in flight. The type of guidance system used in a missile varies according to the desired mission, the size of the missile, and the technology available. Guidance system mechanisms fall into several categories. One type (heat-seeking, radar-seeking) locks the missile directly onto the target. A second (inertial) maintains the missile on a preprogrammed course by measuring and correcting its deviation from that course. A third (satellite, terrain-following) steers the missile by reference to an outside source.

Hot launch. A launch in which a missile is propelled out of its silo by its own exhaust gasses. Compare to "cold launch."

ICBM. Acronym for "intercontinental ballistic missile": a missile with a range greater than 5,500 kilometers (3,300 miles).

IRBM. Acronym for "intermediate range ballistic missile": a missile with a range between 3,000 and 5,500 kilometers (1,800 and 3,300 miles).

Kiloton (KT). A unit used to measure the explosive power, or "yield," of nuclear weapons, equivalent to 1,000 tons of the conventional explosive TNT. Compare to "megaton."

Launch vehicle. A rocket-powered vehicle, steered by a guidance system, that is designed to carry a satellite or spacecraft from Earth's surface into space.

Megaton (MT). A unit used to measure the explosive power, or "yield," of nuclear weapons, equivalent to 1 million tons (1,000 kilotons) of the conventional explosive TNT.

MIRV. Acronym for "multiple, independently targeted reentry vehicle." A single missile carrying MIRVs can strike multiple targets. The acronym is also used as an adjective ("The treaty limits MIRVed ICBMs") and as a verb ("If Russia also chooses to MIRV its SLBMs, the treaty will be in jeopardy").

Missile. A rocket-powered vehicle steered by a guidance system and used as a weapon. Compare to "launch vehicle" and "rocket."

Oxidizer. One of the two principal components of rocket propellant (the other is fuel). Oxidizer supplies the oxygen necessary for the fuel to burn, which enables rockets to work in a vacuum. Jets and other "air-breathing" engines use the atmosphere as a source of oxygen.

Payload. The "discretionary" cargo that a rocket-powered vehicle carries or, more specifically, the weight of cargo it *can* carry. The payload of a missile—the total weight of guidance system and warhead that it can carry—is often called its "throw-weight." Payload is a function of thrust, and the payload of a given missile or launch vehicle varies according to the range or altitude desired. Roughly speaking, the longer the range or the higher the orbit, the lower the payload that can be carried there.

Propellant. An umbrella term for the fuel and oxidizer that power a rocket. Propellants may be solid (in which case the fuel and oxidizer are mixed before being placed in the rocket casing) or liquid (in which case they are carried separately and mixed in the combustion chamber just before ignition).

Range. 1. The maximum horizontal distance that a given rocket or missile can travel; a function of thrust and payload. 2. An uninhabited area over which rockets and missiles are fired for testing and research, such as the White Sands Missile Range in New Mexico. The expression "X miles downrange," heard in broadcasts of NASA launches, is derived from this use of the word.

Reentry vehicle. The separable nose section of a ballistic missile, designed to shield the warhead (and make it easier to steer) as it falls toward the target. See also "MIRV."

Rocket. 1. A machine that is accelerated by accelerating a stream of particles (the working fluid) in the opposite direction. To date, nearly all rockets have used the hot gasses produced by burning propellants as a working fluid. See also "rocket motor" and "thruster." 2. A rocket-powered device without a guidance system, designed to be used as a weapon. Compare to "missile."

Rocket motor. A self-contained rocket designed to be installed as a propulsion system for a vehicle such as an airplane, spacecraft, or missile.

SAM. Acronym for surface-to-air missile: a missile fired from the ground at enemy aircraft. There are also air-to-air missiles (AAMs), air-to-surface missiles (ASMs), and surface-to-surface missiles (SSMs), but SAM is the only one of the four acronyms to become a word in its own right.

SLBM. Acronym for submarine-launched ballistic missile: a missile designed to be carried aboard and launched from specially designed submarines.

Sounding rocket. A rocket-powered vehicle designed to carry small payloads such as cameras or scientific instruments into the upper layers of the atmosphere.

Spacecraft. A vehicle with guidance- and attitude-control systems, designed to operate in space. The category encompasses vehicles with and without human crews, vehicles capable of leaving Earth under their own power (like the space shuttle), and vehicles that must be lifted to orbit aboard launch vehicles (like Apollo, Soyuz, or planetary probes).

Specific impulse. The amount of thrust generated by 1 pound (or 1 kilogram) of fuel in 1 second; a measure of the efficiency of a rocket motor.

Thrust. The force that moves a rocket forward, usually expressed in pounds in the United States and in newtons elsewhere.

Thruster. A term sometimes applied to rockets that do not use the combustion of propellants to produce a working fluid. Small thrusters that use compressed gas as a working fluid have been used to control spacecraft and high-altitude research aircraft since the 1950s. Larger "ion thrusters" that use electricity to accelerate a stream of electrically charged atoms are now being developed as propulsion systems.

Warhead. The destructive part of a missile's payload, consisting of the weapon itself (chemical explosives, nuclear explosives, toxic chemicals, or biological agents) and their associated fuses, triggers, and dispersal mechanisms.

Working fluid. The material, usually a gas, that a rocket accelerates in order to produce thrust. Most rockets use the gasses produced by the burning of their propellants as a working fluid.

Further Reading

GENERAL

Burrows, William E. *This New Ocean: The Story of the First Space Age*. New York: Random House, 1998.

Crouch, Tom D. *Aiming for the Stars: The Dreamers and Doers of the Space Age*. Washington, DC: Smithsonian Institution Press, 1999.

Gunston, Bill. *The Illustrated Encyclopedia of the World's Rockets and Guided Missiles*. London: Salamander Books, 1993.

Heppenheimer, T. A. *Countdown: A History of Space Travel*. New York: John Wiley, 1997.

Launius, Roger D., and Dennis R. Jenkins, eds. *To Reach the High Frontier: A History of U.S. Launch Vehicles*. Lexington: University Press of Kentucky, 2002.

Siddiqi, Asif A. *Sputnik and the Soviet Space Challenge* and *The Soviet Challenge to Apollo*. Gainesville: University of Florida Press, 2003. Originally published in one volume by NASA, as *Challenge to Apollo* (2001).

Wade, Mark. *Encyclopedia Astronautica*. http://www.astronautix.com.

CHAPTER 2: THE AGE OF BLACK POWDER

Crosby, Alfred W. *Throwing Fire: Projectile Technology through History*. New York: Cambridge University Press, 2002.

Von Braun, Wernher, and Frederick I. Ordway III. *The Rockets' Red Glare*. Garden City, NY: Anchor Press/Doubleday, 1976.

Winter, Frank H. *The First Golden Age of Rocketry: The Congreve and Hale Rockets of the Nineteenth Century*. Washington, DC: Smithsonian Institution Press, 1990.

CHAPTER 3: THE BIRTH OF MODERN ROCKETRY

Clary, David A. *Rocket Man: Robert H. Goddard and the Birth of the Space Age*. New York: Hyperion Press, 2003.

Emme, Eugene M. *The History of Rocket Technology: Essays on Research, Development, and Utility*. Detroit, MI: Wayne State University Press, 1964.

Ward, Bob. *Mr. Space: The Life of Wernher von Braun*. Washington, DC: Smithsonian Institution Press, 2004.

Winter, Frank. *Prelude to the Space Age: The Rocket Societies, 1924–1940*. Washington, DC: Smithsonian Institution Press, 1983.

CHAPTER 4: ROCKETS IN WORLD WAR II

Brown, Eric. "Messerschmitt Me-163." In *Wings of the Luftwaffe*, ed. William Green and Gordon Swanborough, 167–176. Garden City, NY: Doubleday, 1978.

Green, William. *Rocket Fighter*. New York: Ballantine Books, 1971.

Hallion, Richard P. *Strike from the Sky: The History of Battlefield Air Attack*. Washington, DC: Smithsonian Institution Press, 1989.

Irving, David. *The War between the Generals*. Harmondsworth, UK: Penguin, 1982.

Mazzara, Andrew F. "Marine Corps Artillery Rockets: Back through the Future." Produced at U.S. Marine Corps Command and Staff College, May 6, 1987. http://www.globalsecurity.org/military/library/report/1987/MAF.htm (accessed March 1, 2004).

Neufeld, Michael J. *The Rocket and the Reich: Peenemünde and the Coming of the Ballistic Missile Era*. New York: Free Press, 1995.

Ziegler, Mano. *Rocket Fighter*. New York: Bantam, 1984.

CHAPTER 5: ROCKETS FOR RESEARCH

DeVorkin, D. H. *Science with a Vengeance: How the Military Created the U.S. Space Sciences after World War II*. Berlin: Springer-Verlag, 1993.

Miller, Jay. *The X-Planes: X-1 to X-31*. Arlington, TX: Aerofax, 1988.

Rosen, Milton. *The Viking Rocket Story.* New York: Harper, 1955.

CHAPTER 6: BALLISTIC MISSILES AND THE COLD WAR

Chang, Iris. *The Thread of the Silkworm.* New York: Basic Books, 1996.

Fitzgerald, Frances. *Way Out There in the Blue: Reagan, Star Wars, and the End of the Cold War.* New York: Simon and Schuster, 2001.

Hall, Rex D., and David J. Shayler. *Rocket Men: Vostok and Voskhod, the First Soviet Manned Spaceflights.* Berlin: Springer-Verlag, 1998.

Miller, David. *The Cold War: A Military History.* New York: St. Martin's, 1998.

Newhouse, John. *War and Peace in the Nuclear Age.* New York: Vintage Books, 1990.

Nolan, Janne E. *Trappings of Power: Ballistic Missiles in the Third World.* New York: Brookings Institution, 1991.

Stumpf, David K., and Jay W. Kelley. *Titan II: A History of a Cold War Missile Program.* Fayetteville: University of Arkansas Press, 2000.

Zaloga, Steven. *Target America: The Soviet Union and the Strategic Arms Race.* Novato, CA: Presidio Press, 1993.

CHAPTER 7: ROCKETS TO THE MOON

Chaikin, Andrew. *A Man on the Moon: The Voyages of the Apollo Astronauts.* New York: Viking/Penguin, 1993.

Gray, Mike. *Angle of Attack: Harrison Storms and the Race to the Moon.* New York: Norton, 1992.

Harford, James. *Korolev: How One Man Masterminded the Soviet Drive to Beat America to the Moon.* New York: Wiley, 1997.

Kelly, Thomas J. *Moon Lander: How We Developed the Apollo Lunar Module.* Washington, DC: Smithsonian Institution Press, 2001.

Lindroos, Marcus, ed. "The Soviet Manned Space Program." 1996. http://www.fas.org/spp/eprint/lindroos_moon1.htm (accessed December 18, 2002).

Murray, Charles, and Catherine Bly Cox. *Apollo: The Race to the Moon.* New York: Simon & Schuster/Touchstone, 1989.

CHAPTER 8: TACTICAL MISSILES IN THE COLD WAR

Hastings, Max, and Simon Jenkins. *The Battle for the Falklands.* New York: Random House, 1984.

Michel, Marshall L. *Clashes: Air Combat over North Vietnam, 1965–1972*. Annapolis, MD: Naval Institute Press, 1997.

Nordeen, Lon O. *Air Warfare in the Missile Age*, 2nd ed. Washington, DC: Smithsonian Institution Press, 2002.

Westrum, Ron. *Sidewinder: Creative Missile Development at China Lake*. Annapolis, MD: Naval Institute Press, 1999.

CHAPTER 9: SPACEFLIGHT BECOMES ROUTINE

Harland, David M. *The Space Shuttle: Roles, Missions, and Accomplishments*. New York: John Wiley, 1998.

Jenkins, Dennis R. *Space Shuttle: The History of the National Space Transportation System*. North Branch, MN: Specialty Press, 2001.

Redfield, Peter. *Space in the Tropics: From Convicts to Rockets in French Guiana*. Berkeley and Los Angeles: University of California Press, 2000.

CHAPTER 10: MISSILES AFTER THE COLD WAR

Atkinson, Rick. *Crusade: The Untold Story of the Persian Gulf War*. Boston: Houghton Mifflin, 1993.

Graham, Bradley. *Hit to Kill: The New Battle over Shielding America from Missile Attack,* revised ed. New York: Public Affairs, 2003.

Hallion, Richard P. *Storm over Iraq: Air Power and the Gulf War*. Washington, DC: Smithsonian Institution Press, 1992.

Index

NOTE: Names of specific *types* of launch vehicles, missiles, rockets, and spacecraft are indexed under those categories.

About the Author

A. BOWDOIN VAN RIPER is Adjunct Professor at Southern Polytechnic State University. He is the author of *Looking Up: Aviation and the Popular Imagination* (2003), *Science in Popular Culture: A Reference Guide* (Greenwood, 2002), and *Men among the Mammoths: Victorian Science and the Discovery of Human Prehistory* (1993).